Anastasiya Shtaltovna

Servicing Transformation

ZEF Development Studies

edited by

Solvay Gerke and Hans-Dieter Evers

Volume 23

Center for Development Research (ZEF)
University of Bonn
www.zef.de

LIT

Anastasiya Shtaltovna

Servicing Transformation

Agricultural Service Organisations and Agrarian Change in Post-Soviet Uzbekistan

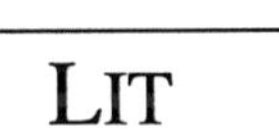

This book was published with the financial support of the Department of Political & Cultural Change of Centre for Development Research, University of Bonn.

This dissertation has been accepted by the Faculty of Agriculture of the Rheinische Friedrich-Wilhelms-Universität Bonn.

First Supervisor: PD Dr. Peter P. Mollinga
Second Supervisor: Prof. Dr. Karin Holm-Müller
Day of Oral Examination: September 18, 2012

Bibliographic information published by the Deutsche Nationalbibliothek
The Deutsche Nationalbibliothek lists this publication in the Deutsche Nationalbibliografie; detailed bibliographic data are available in the Internet at http://dnb.d-nb.de.

ISBN 978-3-643-90358-7

Zugl.: Bonn, Univ., Diss., 2012
A catalogue record for this book is available from the British Library

Klosbachstr. 107
CH-8032 Zürich
Tel. +41 (0) 44-251 75 05
Fax +41 (0) 44-251 75 06
E-Mail: zuerich@lit-verlag.ch
http://www.lit-verlag.ch

LIT VERLAG Dr. W. Hopf
Berlin 2013
Fresnostr. 2
D-48159 Münster
Tel. +49 (0) 2 51-62 03 20
Fax +49 (0) 2 51-23 19 72
E-Mail: lit@lit-verlag.de
http://www.lit-verlag.de

Distribution:

In Germany: LIT Verlag Fresnostr. 2, D-48159 Münster
Tel. +49 (0) 2 51-620 32 22, Fax +49 (0) 2 51-922 60 99, E-mail: vertrieb@lit-verlag.de

In Austria: Medienlogistik Pichler-ÖBZ, e-mail: mlo@medien-logistik.at
In Switzerland: B + M Buch- und Medienvertrieb, e-mail: order@buch-medien.ch
In the UK: Global Book Marketing, e-mail: mo@centralbooks.com
In North America: International Specialized Book Services, e-mail: orders@isbs.com
e-books are available at www.litwebshop.de

Dedicated to agricultural service providers
in Khorezm, Uzbekistan

ABSTRACT

In this thesis, I examine the role of agricultural service organisations in the agrarian and rural transformation processes in Uzbekistan. The two central research questions I address are: how have the social relations of production between the state and agricultural producers, as exemplified by agricultural service providers, changed during transition period? How have agricultural service providers evolved and what is the role they play during the process of agrarian change in Uzbekistan?

Uzbekistan has experienced a chain of agricultural reforms since 1991. The Uzbek government maintains strong control over its agricultural production to ensure food security, employment and economic growth. Scientific literature and reports by international organisations suggest growing poverty and lack of jobs in the rural areas, as a result of which many people move out to Russia and other countries.

Agricultural service organisations (AGSOs) play a crucial role in agricultural production in offering agricultural inputs and sales, financial and mechanisation services. Moreover, they provide employment opportunity in the rural areas. Originally, agricultural service organisations were established to serve state collective farms during Soviet times, and were centrally managed. Due to agricultural reforms that led to the creation of individual farms, AGSOs have moved from providing services for a few state farms during the Soviet planned agricultural system to providing services to a much larger contingent of individual farmers for the current procurement system. The focus of the current agricultural system in Uzbekistan is on cotton and wheat production. Despite the reforms, the development of agricultural service organisations is inadequate, and many of them are on the brink of bankruptcy. This is caused by the end of state supply after the demise of the Soviet Union, debts of farmers, poor input provision, inability to renew hardware such as agricultural machinery and equipment, under-developed market infrastructure in rural areas, mismanagement of agricultural service organisations' resources, lack of access to credit facilities, and a lack of knowledge on how to work under a changing and market-oriented environment.

The data were collected through various qualitative methods in 2009 and 2010 in altogether 11 months of fieldwork. Amongst others, a case study approach was taken to study in detail particularly a Machine-Tractor-Park, the Fertiliser

Company, a private Bio-laboratory and a private international chemical pest control organisation.

Agricultural service providers offered a unique window into the agrarian change process in Uzbekistan. The results show that agricultural service organisations apart from providing services (i.e. provision of fertilisers, seeds, machinery), fulfil many other socio-political functions to society. These include various tasks requested by the state administration in regard to the state procurement system, such as providing machinery to state farmers, and participating in numerous meetings arranged by the state administration. Agricultural service organisations fulfil the private tasks of bureaucrats and local elites as remnants from Soviet times. Agricultural service organisations also act as a social security net for their personnel. The role of agricultural service organisations during the agrarian change was uncovered by developing a heuristic tool in the form of a 'relational typology of agricultural service organisations'. My analysis reveals that agrarian transformation has produced three types of service organization, which differ in their importance to the central government as an income source, their importance towards the smooth operations of the state procurement system, and their capacity to move faster or slower (or even move at all) towards a market economy.

ZUSAMMENFASSUNG

In dieser Doktorarbeit untersuche ich die Rolle von landwirtschaftlichen Dienstanbietern, insbesondere eines Maschinen-Traktor-Parks, einer Düngemittelfabrik und eines Biolabors, im Rahmen landwirtschaftlicher Transformationsprozesse in Usbekistan. Hierbei stehen die folgenden zwei Forschungsfragen im Vordergrund:

Wie verändern sich die sozialen Beziehungen zwischen Staat und landwirtschaftlichen Betrieben innerhalb des Produktionsprozesses als Teil momentaner Transformationsprozesse?

Wie entwickeln sich landwirtschaftliche Dienstleister und welche Rolle spielen sie im landwirtschaftlichen Transformationsprozess in Usbekistan?

Usbekistan hat seit 1991 eine Reihe von landwirtschaftlichen Reformen erlebt. Die usbekische Regierung behält weitreichende Kontrolle über die landwirtschaftliche Produktion um Ernährungssicherung, Beschäftigung und wirtschaftliches Wachstum sicherzustellen. Wissenschaftliche Veröffentlichungen, sowie Berichte von internationalen Organisationen deuten darauf hin, dass Armut und Arbeitslosigkeit in ländlichen Gebieten zunehmen. Als Folge davon ziehen viele Bewohner der ländlichen Gebiete auf der Suche nach Arbeit nach Russland oder in andere Länder.

Landwirtschaftliche Dienstanbieter spielen eine entscheidende Rolle in der landwirtschaftlichen Produktion bei der Bereitstellung landwirtschaftlicher Produktionsmittel, beim Verkauf, und bei Finanz- und Versicherungsdienstleistungen. Darüber hinaus bieten sie Arbeitsplätze in ländlichen Gebieten. Ursprünglich wurden die Dienstanbieter für die staatlichen Kolchosen und Sowchosen der Sowjetunion gegründet und zentral geführt. Mit dem Entstehen einer Vielzahl von individuellen Betrieben im Rahmen landwirtschaftlicher Reformen haben sich die Dienstanbieter verändert. Sie sind nicht mehr Dienstleister für wenige Kolchosen im planwirtschaftlichen sowjetischen System, sondern für eine große Zahl von individuellen Betrieben im gegenwärtigen staatlichen Beschaffungssystem, mit einem Fokus auf der Produktion von Baumwolle und Weizen.

Trotz der landwirtschaftlichen Reformen wurde die Entwicklung der Dienstanbieter unterbrochen und viele von ihnen stehen am Rande eines Bankrotts. Gründe hierfür sind zum Beispiel: wegfallende Unterstützung nach dem Ende der Sowjetunion, Schulden der Bauern, die Unfähigkeit physisches Kapital wie landwirtschaftliche Maschinen und Ausrüstung zu erneuern, unterentwickelte Infrastruktur der Märkte in ländlichen Gebieten, das Mismanagement der Ressourcen der landwirtschaftlichen Dienstanbieter, fehlender Zugang zu Krediten, und fehlendes Wissen darüber wie sie in einem sich wandelnden und marktorientierten Umfeld arbeiten können.

Methodisch basiert die Arbeit auf vornehmlich qualitativen Daten, die im Zeitraum von elf Monaten in den Jahren 2009 und 2010 erhoben wurden. Unter anderem wurde eine Fallstudie als Herangehensweise gewählt, um einen Maschinen-Traktor-Park, eine Düngemittelfabrik und ein Biolabors im Detail zu untersuchen.

Die Agrardienstleister boten einen einmaligen Einblick in den landwirtschaftlichen Transformationsprozess in Usbekistan. Die Ergebnisse zeigen, dass die landwirtschaftlichen Dienstanbieter außer dem Anbieten von Diensten (Bereitstellung von Düngemitteln, Saatgut und Maschinen) viele weitere sozialpolitische Funktionen für die Gesellschaft erfüllen. Sie übernehmen verschiedene Aufgaben die von der Administration im Rahmen des staatlichen Beschaffungssystems verlangt werden; sie erfüllen private Aufgaben von Bürokraten und lokalen Eliten und sie stellen soziale Absicherung für ihr Personal dar. Die Rolle der Agrardienstleister wurde aufgedeckt mithilfe des entwickelten heuristischen Tools „relational typology of agricultural service organization". Die Analyse zeigt, dass die landwirtschaftliche Transformation drei Typen von Agrardienstanbietern hervorgebracht hat. Diese unterscheiden sich bezüglich ihrer Bedeutung als Einkommensquelle für die Zentralregierung, für das staatliche Beschaffungssystem, sowie ihrer Fähigkeit sich in Richtung Marktwirtschaft zu bewegen.

TABLE OF CONTENTS

FIGURES

TABLES

GLOSSARY OF UZBEK, RUSSIAN AND LATIN TERMS

Dekhan farm	(Uzb.)	A household/subsistence production or peasant farmers with approx. 0.25 ha of land
Hokim	(Uzb.)	State administration body or district governor
Hokimyat	(Uzb.)	State administration on the local, district and regional levels
Kolkhoz	(Rus.)	The large collective farm
Komsomol	(Rus.)	Youth division of the Communist Party of the Soviet Union. The Komsomol in its earliest form was established in urban centres in 1918
Militia	(Rus.)	An equivalent to the police service
Shirkat	(Uzb.)	Joint stock companies, reorganised collective and state-owned farms during the independence period
Sovkhoz	(Rus.)	The large state-owned farms
Uzhimprom	(Rus.)	Uzbekistan chemical production or State Joint Stock Company 'Uzkimyosanoat'
Uzneftegas	(Rus.)	Uzbekistan Fuel and Gas production company

Bracon	(Latin)	It is a simplified version of Grabrobrakon *Juglandis Bracon hebetor*. It is an insect that can be bred in the bio-laboratory. The insects are released over the cotton fields. Bracon puts its eggs into the body of larva (or worm). When the bracon's eggs develop in the body of larva, the larvas are destroyed. Thus, bracon protects cotton from worms.
Udarnik 90 days of cotton	(Rus.)	Shock worker 90 days of Cotton
Uzavtodaryatrans	(Rus)	Uzbek Darya transport company
Selector	(Rus.)	A system of phone connection for swift communication of phones in different locations with the central one

ABBREVIATIONS AND ACRONYMS

AGSOs	Agricultural service organisations or providers
AMTP	Alternative machine tractor park; a reorganised machine park of the *kolkhozes* and *sovkhozes*
Bio-lab	Bio-laboratory
F. Co	Fertiliser Company
MTP	Machine Tractor Park
MAWR	Ministry of Agriculture and Water Resources
P.P.	Plant Protection (organisation)

CHAPTER 1: INTRODUCTION

1.1 Problem definition and research questions

Uzbekistan has gone through 60 years of Soviet collectivised and planned agriculture and 20 years of post-Soviet agriculture. The latter included three major reforms, in which aspects of the Soviet and post-Soviet systems were intermixed; Soviet history continues to shape the social, political and economic landscapes of Uzbekistan (Trevisani 2009; Shtaltovna et al. 2011). To ensure food security, rural employment as well as profits from cotton production, the government maintains strong control over agricultural production (Kandiyoti 2003; Spoor 2004; Khan 2005; Wehrheim 2008). Cotton - as in Soviet times - remains the 'white gold' of Uzbekistan. During the Soviet era, Uzbekistan produced two-thirds of world cotton production (Rumer 1989); Today, Uzbekistan continues to be one of the world's major cotton producers (Bremer Cotton Report 2008). Since the country gained independence, cotton has remained the most important foreign exchange income for the national economy, contrary to a significant reduction on the contribution of agriculture to the GDP from 37% in 1991 to around 18% in 2009 (Khan 2005; Spoor 2006; Lerman 2008; Veldwisch 2008; World Bank 2011). Currently, a state procurement system is in place where farmers fulfil orders from the government, in particular for crops like cotton and wheat.

Van den Ban and Hawkins noted previously that the most difficult problems former communist countries face are the transformation of workers of former cooperative and state farms into private entrepreneurs, and the building of input supply infrastructure, marketing and extension that are necessary to make private farming viable (1996). Some reasons for these problems in Uzbekistan, and similarly in Kazakhstan, include a lack of experience of a changing environment, financial constraints, a lack of machinery, bureaucratic obstacles, and unfair redistribution of land and farm assets during privatization (Gray 2000; Peabody et al. 2000; Shreeves 2002 in Toleubayev 2009).

In post-Soviet Uzbekistan, well-functioning agricultural service organisations (AGSOs) are necessary for rural transformation and a successful transition to commercial agriculture. These service providers play a crucial role by providing agricultural inputs, sales organisations, financial and insurance services (Shtaltovna 2011). Moreover, AGSOs create working opportunities in rural and regional areas that are important in preventing mass migration from the rural areas to big cities and abroad. From the perspective of the state, AGSOs are not only a form or instrument of regulation; these organisations also act as incentive creation structures intended to facilitate and enhance agricultural production activities, as presented in the following chapters.

Originally, agricultural service providers were established during Soviet times to serve state collective farms. Due to the chain of agricultural reforms, AGSOs have moved from being centrally managed and providing services to state and collective farms (sovkhozes & kolkhozes) during the Soviet planned agricultural system to servicing a much larger contingent of individualised farmers during the current procurement system (with the focus on cotton and wheat production) in Uzbekistan. However, AGSOs have not managed to evolve successfully, and many of them are on the brink of bankruptcy (Shtaltovna et al. 2011). This is caused by factors such as a ceasing of state supply, debts of farmers, poor condition of the service and input provision, inability to renew hardware (such as agricultural machinery and equipment), poor education, under-developed market infrastructure in the rural sector, abuse by local elites and people working in the bureaucracy of AGSOs' resources, lack of access to bank credit facilities, and a lack of knowledge of how to work under a changing and more market-oriented environment (Shtaltovna 2011).

Transition and agrarian change consist of the process in which the features and activities of Soviet period agricultural service organisations take on new shapes in post-Independence processes. Not only do the functions and activities of organisations change, but new organisations appear while existing ones disappear. It is a dynamic process that my thesis looks at and clarifies: how these changes happen, what they mean, and finally, where this may lead.

There has not been a detailed study on changes in agricultural service provision in any of the post-Soviet states since the demise of the Soviet Union. This thesis contributes to the lack of knowledge in this area by analysing agrarian change and rural transformation processes in Uzbekistan through the prism of AGSOs located in Khorezm province. The two main questions addressed in this study are 1) How have the social relations of production between state and agricultural producers exemplified by AGSOs changed during the transition period? 2) How have AGSOs evolved and what is the role they play during the process of agrarian change in Uzbekistan?

The following sub-questions were used as guidance:

- What kinds of AGSOs have emerged during twenty years of reform in agriculture?

- What makes an organisation good and successful during the transition process?
- Which organisations are likely to survive the transitional process?
- Which linkages are likely to prove essential for survival under conditions of transition?
- How can organisations, as organisations, be improved in order to support agricultural/regional development?
- Who are the driving forces for the reinvention of AGSOs during the transition period?
- What is the role AGSOs play for the central government during the transition period?
- What is the role AGSOs play for clients during the transition period?
- What is the role AGSOs play for their employees during the transitional period?
- What is the role AGSOs play for the state elites during the transition period?
- What do we learn, by studying AGSOs, about the 'authoritarian state' in the transition environment?

To answer these sub-questions, I have analysed the internal and external organisational relationships of three AGSOs to determine the space within which they are allowed to reinvent themselves. Specifically, I have carried out in-depth studies of a Machine Tractor Park, a Fertilizer company and a Bio-laboratory in Khorezm province and a careful mapping of their local, regional and national environments. The study of AGSOs offers a unique window onto the agrarian change and rural transformation processes in Uzbekistan. Socially, the results of this research, namely the detailed description and in depth analysis of agricultural service providers will be useful for policy recommendations for restructuring these organisations.

1.2 Setting and research background

Uzbekistan has a land area of 447,400 km². It is a resource rich, doubly-landlocked country, strategically located in the heart of Central Asia (Figure 1). It borders Afghanistan, Kazakhstan, the Kyrgyz Republic, Tajikistan and Turkmenistan. Uzbekistan accounts for one-third of the Central Asian region's population with 27 million inhabitants. The agricultural sector employs around 30 percent of the Uzbek population (Hough and Fainsod 1979).

Khorezm province is located in the irrigated lowlands of the Amu Darya River, which is one of the two major rivers that used to flow into the Aral Sea. There are 1.3 million inhabitants in Khorezm and about 70 percent live in rural areas (Wehrheim 2008). Agriculture in Khorezm region is the main sector of employment. The majority of inhabitants are involved in the agricultural sector. At present cotton, wheat, rice, fodder maize, fruits and vegetables are the main agricultural crops of Khorezm.

Khorezm is used as a case study of the transitional environment of Central Asia and assessing the evolultion of agricultural service organisations.

Figure 1: Khorezm province on the map of Uzbekistan, Central Asia. Source: University of Texas (2012). To illustrate the data collection site, the map was modified by A. Lee, GIS laboratory, ZEF/UNESCO project, Urgench 2012

This research was conducted within the frame of the ten year long interdisciplinary Zentrum für Entwicklungsforschung (ZEF)/ United Nations Educational, Scientific and Cultural Organisation (UNESCO) project 'Economic and Ecological Restructuring of Land- and Water Use in the Region Khorezm (Uzbekistan): A Pilot Project in Development Research' funded by the Federal Ministry of Education and Research (BMBF), Germany (Project Number 339970D) and implemented by the Center for Development Research /ZEF, University of Bonn, Germany. The research builds on, while at the same time contributing to, a large stock of primary and secondary data on the many facets and faces of everyday forms of agrarian change in post-Soviet Uzbekistan. In the course of the project, research in various disciplines was carried out gathering quantitative and qualitative evidence on the economic, social and ecological issues prevalent in the study region.

As part of Phase III, the final stage of the ZEF/UNESCO project, this thesis investigated agricultural service providers in Khorezm. During this phase, alternative policies, institutional arrangements, and integrated technologies were explored and tested from 2007 to 2011. In the final transition period, the local partner – the University of Urgench – will take the lead in continuing with the research, implementation, and adoption of technologies commenced during the project's lifetime[1].

1.3 Significance of the study and main findings

The specific contribution that the analysis of AGSOs' functioning can make is that it provides a close look at the changes in relationships between the state and agricultural producers, a core aspect of the 'reform' of a 5rganizatio and 5rganizationa agrarian system. The study thus contributes to ongoing debates on agrarian change and rural transformation processes in Uzbekistan. This study is important as it investigates the field of agricultural service provision in Uzbekistan, which has not been comprehensively studied yet in either Uzbekistan or other former Soviet countries. In-depth case studies give a thorough understanding of the problems the organisations face during the ongoing agrarian and transition process. Therefore, the results of this study can be used for the restructuring of agricultural service provision in Uzbekistan.

Investigating agricultural service providers twenty years after the demise of the Soviet Union gives us an understanding about certain processes of transition that have taken place in Uzbekistan. Looking at AGSOs provides a window to how the state and service providers negotiate, how different objectives of people collide and are resolved. Agricultural service providers are one of the arenas

---

1 For more detailed account of the Project and project publications please visit a website http://www.khorezm.zef.de/

where the Soviet legacy is being reshaped under new conditions. The main findings of the research are the following:

1 **Centralised state rule remains dominant in Uzbekistan, and the hands of agricultural producers are tied to the state procurement system despite the individualization of farms and claims that the country is moving towards a market economy. Framed by ongoing agrarian change and the state procurement system, many service providers and most state organisations at the regional and district levels can be regarded as state instruments of regulation and incentive structure creation intended to facilitate and enhance agricultural production activities.** The empirical evidence showed that the relationships with the state proved to be very strong and have a strong impact upon functioning of agricultural service organisations. The strong position of the state in present day agricultural politics also has an impact upon service providers' relationships with farmers. Under the present state procurement system in agriculture, AGSOs are the interface between the state and farmers, with a very strong influence of the state. Therefore, it makes the system is an extreme form of Technical and Administrative Task Environment (TATE) (Benvenuti 1975a).

2 **Service providers provide a basis for obtaining private and collective goods for their employees and for other state actors involved in state agricultural production during the agrarian change in Uzbekistan.** Service organisations not only provide agricultural services (e.g. provision of fertilisers, seeds, and machinery) to the state procurement system, but also fulfil many other socio-political functions in society. For example, they fulfill all assignments by the state administration in regard to the state procurement system even those, which are not directly related to their job profiles. They serve as knowledge carriers for farmers and other actors currently involved in agricultural production, as well as act as a social security net for its staff and state bureaucrats during and due to the transition. As a result of the multiple roles that agricultural service providers play in the society during the transitions, they are torn between different expectations by the state and society and fail to evolve as individual units.

3 **The situation of service providers, similarly to farmers, will not change fundamentally (either for the better or for the worse) as long as the state procurement system is in place.** The reason why many service providers cannot really move forward or evolve is that they were established to serve collective and state farms during the Soviet times and

now continue to serve state agricultural aims. Therefore, they have no experience in running a business independently since they have never experienced a market environment. Being organized in this particular way, the state agricultural system contributes to the wellbeing of the Uzbek centre. In this way it keeps many elements of the Soviet Union system in play, thus the country cannot extricate itself from many rules, especially informal ones, which worked during the Soviet times. Therefore, the majority of the service providers cannot move ahead and develop in a market economy.

4 **Notwithstanding overwhelming efforts of central state control and regulation, there are dynamics on the ground towards a market economy. During the transition (so far) three types of service providers have emerged.** The role and the nature of AGSOs during the agrarian change were examined with the help of 'Relational typology of agricultural service organisations'. Relational typology has shown to be a useful technique to understand AGSOs in the period of transition. Relational typology helped to illustrate the nature of established relations as well as dependencies between the state organisations, agricultural service providers and farmers. Apart from being a useful tool to study service providers under transition, relational typology defined three types of agricultural service organisations that exist in the current transition process in the context of Uzbekistan. AGSOs were classified into three categories according to the following criteria:

a. Importance of the 7rganization to the central government as an income source;

b. Importance of the 7rganization to the state procurement system;

c. The organisations' capacity to move faster or slower (or move at all) towards a market economy;

d. The organisations' relationships established with the state local and regional administration;

e. The organisations' relationships established with the state elites and bureaucrats and their apex organisations;

f. The organisations' relationships with their clients;

g. The degree of reinvention and diversification of the 7rganization during and due to the process of agrarian change.

5 **The transformation of the Soviet state apparatus**: Research on AGSOs provided an opportunity to further knowledge on many state organisations that remain today in Uzbekistan since the end of the Soviet Union. Poor state funding and the lack of necessity to sustain this heavy state apparatus has led these organisations to undergo changes, i.e. some are dismantled, some of them run into bankruptcy, some become reorganized, merged and remerged with other institutions. Some examples of such organisations are the apex organisations of the studied AGSOs. MTP Union and Plant Protection, the apex organisations of the Machine Tractor Park (MTP) and bio-labs, respectively, are not needed for the present day agriculture and they are not supported by the central government; both apex organisations are similarly more a burden than of help to their daughter organisations. MTP Union and Plant Protection impose their cooperation on the MTP and bio-labs and thus try to survive at the expense of their daughter organisations. The interaction between the Fertiliser Company (F. Co) and its apex 8rganization is different. The relationship is guided by the central state's interests in agriculture and chemical industries: the apex 8rganization tries to prevent the supply of fertilisers to farmers for free; is also tries to prevent 8rganizational abuse at the local level. The dynamics that take place between AGSOs and their apex organisations inform us of the nature of structural system 8rganizational.

The nature of an authoritarian state as a part of agrarian change: The Uzbek state has different facets and there are different interests in the agrarian change process. There are contradictions and different forces within the Uzbek state. Both are seen in AGSOs and shape what AGSOs are and how they can develop as illustrated by the example of the apex 8rganization of the Fertiliser Company, the *Uzhimprom*. On one hand, the *Uzhimprom*, being part of the state, has to pursue the central government's interest in cotton growing. On the other hand, the *Uzhimprom* stands for the interests in receiving returns from 8rganizati production. An empirical instance where the apex 8rganization assists the regional and local levels of the F. Co to cope with the state abuse of their services is also an example of tensions or contradictions in the Uzbek agrarian transformation process. Tension between prolonging state controlled agriculture and the 'unleashing' of the forces of the market is present within the state system because of the opposing goals of the state organisations. Another example of multiple interests of the state administration workers is with local level state representatives. Local elites and bureaucrats have political and regulating roles, and they can exert power at their local level beyond their formal administrative mandates. While fulfilling their functional and regulative roles, they pursue private strategies and interests at the same time. For example, *hokims* have acquired more space to manoeuvre in the present system than in Soviet times, but have fewer resources. The implementation of state order agriculture together with the 9rganizational structure (path dependency) provides a power source for regional governors. They are subordinate to the central authority in Tashkent, which is the ruling state-ordered agriculture, but they are given local power to act freely and extract personal benefits. They need the mandate from the Tashkent, but they also want to keep Tashkent at a distance for their own purposes. This situation is somewhat different from the Soviet situation. Another example of contradiction within the Uzbek state is limited cash in the agricultural economy. By giving little support for the bio-labs and the Machine Tractor Parks' development, the central government still expects them to fully support state procurement system. The process of finding one's way through the upheavals of transition is presented in the empirical cases and 9rganizational descriptions. The structure at rule in Uzbekistan in my study shows a broader perspective on authoritarian state.

1.4 Thesis outline

The thesis is divided into eight chapters. This **first chapter** has introduced the subject of the study, problem definition and research questions, significance of the study and the main findings. The chapter closes with the thesis' outline.

The **second chapter** offers the conceptual framework of the study. I discuss the rationale of the studied actors' 9obilize under the present day agrarian change. A discussion of behavioural rationale mobilises theories such as patron-clients relations, formal-informal rules, state-periphery discussion, and agency of leaders during the transition period. These theories help to understand the 9obilize of the studied actors under the transition process. Moreover, in this chapter I introduce the setting of the research and methodology. I explore AGSOs by looking at inner 9obilized9on9l relations as well as relations with their outer environment through a survey with farmers, interviews and internships in the Khorezm province.

Chapter three sets the stage of the research. This thesis aims at contributing to our understanding of agrarian change and rural transformation processes in Central Asia. Therefore, in this chapter I engage with conceptual and empirical literature on agrarian change and rural transformation processes, Soviet legacies and political economy, largely with the focus on agriculture in countries belonging to the former Soviet Union. I take a closer look at the agrarian change process in Uzbekistan starting from the initiation of 9obilized9on9 agriculture in the 1930s up to the present-day agriculture. In the process of agrarian change in Uzbekistan, I particularly focus on the role of AGSOs, how they have changed along the process. Based on the conducted farmers' survey, as well as in-depth study of three AGSOs, I present an overview of services available in agriculture, as well as gaps in service provision and problems encountered by farmers while receiving services for different types of agricultural production. The functioning of AGSOs is embedded in the ongoing transition process and here I present issues that significantly impact the daily work of AGSOs and other actors involved in agriculture. As an annex to this chapter, I present a table illustrating changes that have taken place in different agricultural service providers through the process of agrarian change starting from Soviet times till present day. This chapter ends with a detailed description of the three case studies of this research.

Chapters four, five, six and seven are the empirical chapters of the thesis. They are structured according to the characteristics and dynamics presented and suggested by the relational typology of agricultural service providers presented in chapter two. In **chapter four**, I explore in-depth the relationships established between the three studied AGSOs and the state administration and also their apex organisations. Being strongly embedded in Soviet working principles, and due to the current state procurement system 10obilized around cotton and wheat growing, the three studied organisations have established a multitude of relationships with the state administration. The chapter discusses different implications of the relationships that state organisations have with AGSOs.

Chapter five is about agricultural service organisations as a space for local elites' and AGSO's employees personal agendas in servicing their livelihood strategies during agrarian change. I look at the relationships that occur between AGSOs and the state (exemplified by the state administration, a tax inspection, a prosecution office, a technical supervision officer and the police) and AGSOs' apex organisations. I also look at the inner 10obilized10on10l dynamics of three studies cases, where the relationships with the local elites have an impact. This chapter discusses how AGSOs play a role in and are a resource for employees' and state bureaucrats' personal 'projects' of securing and improving their livelihoods during the process of transition. This kind of exploration gives us an understanding of how AGSOs, the state apparatus and the apex 10obilized10on of AGSOs change due to the transition process.

In **Chapter six**, I explore the relationships established between AGSOs and farmers. In particular, this chapter explores the differences established between three studied service providers and different types of farmers (mainly those working for the state procurement system and those growing other kinds of commercial crops). The data presented on how AGSOs and farmers adapt from the Soviet system to the ongoing period bring further evidence to agrarian change and rural transformation process as well contributes to the relational typology of AGSOs.

Chapter seven looks at AGSOs' reinvention and diversification strategies. In each AGSO, the strategies are driven by different motives and actors. I determine the strategies undertaken by AGSOs in order to survive during the transition period. I look at qualities important for the service providing organisations to sustain and move through a multitude of odds with the state, which disturb the work of their 11obilized11on, and thus to move through the transition period.

The thesis ends with the research findings and policy recommendations in **chapter eight.** I summarise the empirical findings of the thesis. Also, I discuss the contributions of my findings to our understanding of agrarian change in Uzbekistan. I reflect on the method employed in this thesis, notably, the heuristic tool 'relational typology of AGSOs' and on conducting research in a politically sensitive environment. Relational typology was developed and 11obilized for uncovering the specificities of interactions and interdependencies amongst the actors in the ongoing transition process in Uzbekistan. The chapter concludes with policy recommendations for the development of AGSOs.

CHAPTER 2: THEORIES, METHODOLOGY AND POST-SOVIET COUNTRIES

2.1 Introduction

As discussed in chapter 1, the two main research questions of this study are: 1) How have the social relations of production between state and agricultural producers exemplified by AGSOs changed in the context of transition? 2) How have AGSOs evolved and what is the role they play during the process of agrarian change in Uzbekistan? To address these questions, this chapter provides the theoretical grounding of my research. In section 2.2, I explain how and why I have approached the exploration of AGSOs in the way I did. The study deploys social theory and political theory to analyse and examine organisations and organisational change in the ongoing transition process. I begin by introducing the concept of Technical and Administrative Task Environment (TATE) in section 2.2. TATE helps to position AGSOs in the agricultural production process in Uzbekistan. The concept provides a foundation for us to understand changes in the social relations of production in Uzbekistan. This concerns, in particular, the relationships between the state and agricultural production units, from the collectivised situation in the Soviet period to the present, and still in flux, semi-privatised individual farming with strong state control. In section 2.3, agency-structure theory is introduced to provide an entry into understanding the dynamics in and of AGSOs in several ways. The theory provides a basis to identify and analyse the mechanisms that link structures and events. Next, in section 2.4, I discuss structural, organisational and individual path dependencies that can be found in AGSOs. It is followed by section 2.5 where I look at agricultural service provision trajectories in other ex-Soviet countries. Section 2.6 discusses the rationales of actors' behaviour under the state procurement system together with a number of additional concepts that will appear in the thesis. Finally, section 2.7 discusses the methodology of the research.

2.2 Changing production relations

There are several theories looking at relations established between agricultural producers and the external environment where agricultural producers function. Two of them are Whatmore and Benvenuti's approaches, which I summarise briefly here. In terms of Whatmore et al.'s (1987) relational typology of British farms, the thesis looks at some of the external relations of production of agricultural production units. In terms of Benvenuti's (1975) rural sociology perspective, the thesis looks at a subset of the Technical and Administrative Task Environment (TATE) of farming. This concept suggests that farmers are related to the outer environment, which includes market agencies and associated institutions, i.e. agricultural industries, banks, trade, extension service and others. I investigated the relationships of agricultural service organisations, which provide services to farmers. Among these external production relations of farms, relationships with AGSOs are one subset.

The literature on agrarian transformation in Central Asia and Uzbekistan is characterised by a focus of analysis on macro indicators of agricultural reform, in terms of performance of the agricultural sector, the structure of cotton and wheat production, the production volumes, the level of employment in agriculture (see Chapter 3). What we know little about are the practicalities of transition: how transition takes place on a day-to-day basis (and whether 'transition' is how those involved experience the process). Moreover, the social relations of production that are located in the heart of the agrarian change process (the dynamics of land tenure, employment relations and others) have barely been studied, with the exceptions of Kandiyoti (2007), Veldwisch (2008), Sehring (2007) and Trevisani (2008).

This thesis will show that AGSOs' dynamics are embedded in the context of transition and agrarian change; the study of agricultural service providers in Khorezm province of Uzbekistan thus provides an entry point for increasing our understanding of the broader process of agrarian change or transition in Uzbekistan, and adds to the stock of empirical knowledge on that process. Therefore, this thesis will contribute to the debate on agrarian change by providing evidence on how the relationships between the state and agricultural production units (as exemplified by AGSOs) have changed from the collectivised situation in the Soviet period to the present, though still in flux, semi-privatised individual farming with strong state control. Furthermore the thesis shows that agrarian change is ongoing.

2.3 Structure, agency and organisational practices

In terms of overall approach of the research, the study situates itself as a sociological study of agrarian transformation. By focusing on the everyday practices of AGSOs as the main research topic, the study adopts a sociological perspective that aims to understand the complex relationship between structure and agency in society as the theoretical key to understanding social transformation

(Giddens 1984; Archer 1995). The relationship can be studied best by investigating the social practices in which agents actively reproduce and transform social structure. The interplay between structures and actors is a key element and thus also a key unit of analysis in critical inquiry into agrarian change (Borras 2011).

Central to my approach is the study of 'dynamics'. Uzbekistan's agrarian and rural structure is characterised by a plethora of 'inherited structures' (from the Soviet period) that at first sight may seem to be pretty rigid and hardened, but within which many things are happening, leading to structural change. The studied system of the agricultural rural economy is in transformation. AGSOs are located at the interface of state and agricultural producers. The AGSOs are inherited structures in transformation. By studying closely the day-to-day events of agricultural service provision, the study traces the processes and mechanisms of this 'structural elaboration' (Archer 1995) of AGSOs[2]. The sociological notion of agency attributes to the individual actor the capacity to process social experience and to devise ways of coping with life, even under extreme forms of coercion. Within the limits of information, uncertainty and other constraints (e.g. physical, normative or politico-economic) that exist, social actors attempt to solve problems, learn how to intervene in the flow of social events around them, and to a degree they monitor their own actions, observing how others react to their behaviour and taking note of the various contingent circumstances (Giddens 1984; Long 2001). According to Giddens, agents can form and implement flexible responses to dynamic situations because agents possess power (1984). Agents have power because they are knowledgeable and because they act through rules and resources, which consequently constitute 'structure'. Therefore, in such a view, the constitution of the agent encompasses structure and thus represents a duality (Llewellyn 2007). An object cannot be studied separately from the environment it is embedded in. Structure and actors (agency) have significant influence upon each other, and each cannot be really understood if studied in isolation (Giddens 1976)[3]. To this end, human agency is not related just to the individual persons who make decisions, but also to organisations such as political parties, state organisations, and the church (Long 2001). Therefore, the concept of agency-structure has been mobilised for

2 On the critical realist ontology of structures, mechanisms and events implied here, see (Sayer 1992)

3 Giddens' structuration theory works around two poles, as Archer's morphogenesis; Archer argues that agency and structure are ontologically distinct, while Giddens sees them as 'two sides of the same coin'. See (Giddens 1984) (Archer 1995) (King 2010)

this study on AGSOs to focus on the social analysis of 'practices', the activities through which knowledgeable and capable actors reproduce and transform structures.

Strathern has pointed to the fact that agency is constructed differently in different cultures, arguing that attributes of agency such as knowledge, power, and prestige are attached differently to the concept of 'person' (1985). Hence, research on social practice should pay attention to this while analysing how power, influence and knowledge may shape responses and strategies of different actors differentially. While conducting such an analysis, Long (2001) stresses the importance of identifying and characterising: different actors' practices, strategies and rationales; the conditions under which they arise; how they interlock; their viability or effectiveness for solving specific problems; and their wider social ramifications.

The investigation of 'organisational practices' informs us of the relationships that AGSOs are part of and constitute. In this thesis, actor-oriented analysis as outlined by Long (2001) is employed to study organisations in Khorezm; the method of analysis suggests a dynamic approach to the understanding of social change that stresses the interplay and mutual determination of 'internal' and 'external' factors and relationships and which recognises the central role played by human action and consciousness. According to Long (2001), the advantage of an actor-oriented approach is that it aims to grasp public concerns and private dilemmas through a systematic ethnographic understanding of the 'social life' of the study – in this case, the social life in and around AGSOs. Essential elements of this ethnographic endeavour focus on the elucidation of internally generated strategies and processes of change.

More specifically, the research can be described as a 'sociology of organisational practice'. For analysing the 'dynamics' of AGSOs, I employ an 'ethnography of organisations' perspective. Ethnography in organisations allows us to learn about the cultures and structures of organisations from the inside out and examines both what people say and what people do (Schwarzman 1993). Organisational ethnography offers a method for research that problematises the ways that individuals and groups constitute organisations (and societies) on a daily interactional basis. Many studies on organisations focused on the inner relations and interpersonal relations in the organisation (Chapple 1941; Chapple 1953; Sayles 1957 in Schwarzman 1993; Burawoy 1979; Bramel and Friend 1981). In my work, I look at the inner interactions within AGSOs as well as relations of AGSOs with the outside world (e.g. the relationships between AGSOs and state organisations, and between AGSOs and their clients). Looking at the inner and outer relations of AGSOs provides a window to the agrarian change process mentioned above and increases our understanding of how organisations manage to sustain themselves in times of change.

2.4 Path dependencies in agricultural service organisations

The Soviet heritage and the extensive agricultural systems built up over decades of central planning created a strong 'path dependency' (Thelen, 1999; Sydow, 2005) of agricultural production originating from the central administrative system of Soviet times. According to the concept of path dependency, it can be argued that historical experiences and policy legacies frame present actions: established patterns of behaviour that already once proved to be successful will be used again to meet new challenges. Without the past as a guide, actors cannot navigate through the present and into the future. It serves as a basis for their institutional preferences and assessments of relative power (Luong 2002: 50). In this work, I would like to distinguish three types of path dependencies. These are dependencies that occur in relation to agrarian structure, organisational structure and personal structure. *Agrarian structure* is the 'external structure' of the state controlled social relations of production in agriculture. The present day system uses the same planned and controlled system in agriculture for retaining managers and workers, which worked during Soviet times. Agrarian structure's path dependency defines the opportunities and constraints for AGSOs to function within it. Chapter 3 will further disucss on agrarian path dependencies. The second type of path dependency during transition refers to *organisational structures*. The inherited internal structure of organisations defines the coping strategies and entrepreneurial strategies – the 'agency' of AGSOs. Organisations, as collective actors, also rely on the past to evaluate the present and future effects of the changes that are taking place around them. By investigating relationships between AGSOs and state organisations, we can see what has been borrowed from the past experiences of the preceding agricultural system and organisational structures of AGSOs in order to fill the gaps of and reinforce the present, still changing, agricultural system (see Chapters 4 and 5). The third type of path dependency I identify is *personal structures*. It is an 'internal structure' of an individual or his/her identity. The inherited personalities are a resource for coping and expanding during agrarian change. The people who work in the present day agriculture have gone through the powerful Soviet institutional system with its precise forms of career control apparatus of personnel management (Kornai 2008, Loung 2002). I will elaborate more on this type of path dependency in Chapter 7 to a large extent and in chapters 4 and 5 to a lesser extent.

2.5 Why other ex-Soviet countries cannot be used as a model for Uzbekistan

The aim of this section is to present the diverse trajectories of service provision in the post-Soviet space, twenty years after the demise of the Soviet Union. I show that the experience of ex-Soviet countries in regard to agricultural service provision is not relevant to Uzbekistan.

I found little scientific literature focused on the investigation of service provision in the post-Soviet period. Most of the information available about AGSOs comes in the form of reports of international organisations like FAO, GCARD, ICARDA, and CACAARI[4]. These studies focus on assessing agricultural sector capabilities as a basis for deciding where to allocate donor money in the scope of regional and rural development. No scientific analysis of AGSOs in the particular conditions of post-Soviet Uzbekistan has been conducted. Therefore, the major part of the overview below was gathered through communications with scientists, NGO workers, representatives of international organisations, extension workers and governors from Ukraine, Tajikistan, Kyrgyzstan, and Kazakhstan[5].

The development of agricultural service provision in ex-Soviet states is similar in the way that they all derived from the previous Soviet system. However, the trajectories of evolution differ due to the political and economic paths each country has taken after the demise of the Soviet Union. Immediately after independence, agricultural services in Ukraine were based on the existing Soviet infrastructure and organisations. Over the years, as state support rapidly declined, many of those institutions have broken down. As Ukraine was openly moving toward a market economy, services and input provision in agriculture were supposed to function fully on a private basis. They worked based on the laws 'On property', 'On entrepreneurship', 'On companies' (Verkhovna Rada 2011), and other laws. For instance, organisations like the Machine-Tractor

4 FAO –Food and Agriculture Organisation, CAC - Central Asia and the Caucasus, GCARD – Global Conferences on Agricultural Research for Development, ICARDA – International Centre for Agricultural Research in the Dry Areas, CACAARI – Central Asia and Caucasus Association for Agricultural Research Institutions

5 Find a list of interviewees in the Annex 3.

Parks and the Fertiliser Company, which are case studies for this thesis (for more details see Section 3.6) existed in Ukraine in Soviet times as well. However they did not last long in Ukraine. Other forms of sales and production of fertilizers occurred on the market very soon and thus, there are diverse providers of those services (Mosiyuk 2001:179). Agricultural inputs are provided by local or foreign producers and available on markets and in private specialized shops. Some examples are that much machinery is imported from Belarus; certain types of pesticides come from multinational companies like Du Pont from China (Interview with Nesterov I., Product Marketing Manager, Ukraine and Moldova CNH Services srl, Representative office in Ukraine). The use of services by different types of producers varies a bit due to their different economic standings. For instance, big agricultural companies buy own agricultural machinery or rent it from firms offering foreign combines. The smaller producers cannot always afford renting foreign machinery and thus face the problem of using obsolete old machinery. Rural households, who grow crops mainly for personal consumption, can buy any pesticide on the markets and in specialized shops. However, purchasing power remains a serious obstacle in obtaining services by rural households and small agricultural producers. If in earlier years of agrarian transformation there were some state support programs for agricultural producers, i.e. subsidised loans and lower or no taxes, such facilities are not there anymore now. Bank loans for agricultural producers are at the rate of 25% annual interest (Zubets 1996; Mosiyuk 2001; Uzbekistan Agroportal 2010; World Bank 2011). As a solution, some fertiliser providers allow agricultural producers to pay for services partially in kind, i.e. wheat (Mosiyuk 2001).

In countries like Tajikistan and Kyrgystan the economic and political situation, together with the mountain geography of the countries, make the functioning of agriculture and service provision challenging. There is hardly any state support for agricultural producers; many inputs in agriculture are supported by international donor organisations like RAS, World Bank, USAID, Swiss Agency for Development and Cooperation, FAO, Helvetas, GIZ and UNDP. International donor support is implemented via NGOs or function on a private basis. In Kyrgystan approximately 90% of inputs are given by donors to farmers for free (interview with Nasyrova A., chairperson of management board of Training and Extension System 'TES centre', Kyrgystan, 2010). The remaining 10% are covered by farmers and service providers. The private sector is active in supplying inputs, seeds, breeds, provision of credits, marketing, and processing. The state is involved when the seeds or fertilizers are imported. Imported seeds are checked and registered by respective state checking organisations and laboratories if there are such. Therefore, service to agricultural producers in

Kyrgystan covers all fields of services and is mainly provided via projects funded by international donors and the private sector, with very little state intervention in this process. Recently the political problems and instability in the region caused by rioting in the city of Osh *between ethnic Kyrgyz and the minority Uzbek population* in June 2010 in Kyrgystan has negatively impacted, basically blocked international donor funding for input supply in Kyrgystan.

Following the demise of the Soviet Union and the economic depression caused by the ensuing civil war, Tajikistan's food production declined drastically (World Bank 2011). In Tajikistan, there were neither state programmes, nor subsidies for the agricultural inputs from the state during the first years of independence. Until 2006, donor organisations[6] distributed inputs to farmers free of charge (interview with Muhiddin Nurmatov, USAID extension consultant, 2012; Petra Geraedts, team leader EU SENAS Tajikistan, Cardno UK, 2010). The distribution of inputs free of charge disturbed the development of the private markets in this area. After 2006 instead of free of charge inputs, with the help of international organisations micro-credit schemes were introduced and mini-agro-shops emerged (Interview with the Suleymanova M., director of Public organisation "SAS Consulting; interview with Mamadshoev A., an agronomist at Mercy Corps, Shahrituz, Tajikistan, 2012). High quality seeds are difficult to find (Interivew with the First Deputy Mayor in charge with agriculture, Pengikent, Tajikistan, April 2012). The problems the farmers face here are access to the irrigation water, fake materials brought from China, which are very difficult to distinguish from real ones. If the product is not fake it costs much more due to the expensive, long-lasting and heavily bureaucratized procedure of testing of the imported new seeds, fertilizers and others. There is contraband of cheaper inputs from other central Asian states, however, the quality is poor. Due to political conflicts between Uzbekistan and Tajikistan, the import procedure of fertilisers is very cumbersome (Interview with Nematov I., the director of NGO 'Bakht', Tajikistan, 2012). Moreover, due to the political instability in Kyrgystan (the recent events in Osh district of Kyrgystan (2010) import of inputs into Tajikistan from the east is blocked.

6 TAFF - The *Tajik* Agricultural Finance Framework, GIZ – German Organisation for International Cooperation, Caritas, Aga Khan Foundation, USAID, US Embassy, Mercy Corps, ACTED – French Agency for Technical Cooperation and Development, PRO-APT - The Productive Agriculture in Tajikistan project, GAA – German Agro Action, Helvetas – Swiss Intercooperation, ICCO – Interchurch Organization for International Cooperation JICA, CESVI .

Furthermore agricultural mechanisation in Tajikistan is very limited. New agricultural machinery is hardly available and what is there is funded by international projects. Otherwise, old machinery of Soviet times and manual labour is used. Bank loans for agricultural machinery are available but at the rate of 36% annual interest (Interview with Nurmanadov R., the director of the NGO 'Jovid', Tajikistan, 2012). A small share of cotton production in Tajikistan is grown under state control (ICG 2012). Howerver, the level of control differs a lot and it is reducing every year (Interview wtih Ghulomhaydarov S., an agricultural specialist at the NGO 'Shifo', Tajikistan, 2012).

In Tajikistan and Kyrgystan, extension services, which are organized as NGOs, connect agricultural producers with micro-credits, agricultural banks, and other credit programs. Also, extension tries to ensure that farmers get higher quality of inputs and receive more information about different supply opportunities. Bringing farmers with input suppliers via extension service showed to work well in Tajikistan and Kyrgystan, thus offering a win-win situation for both (LBL 2002; GIZ/ZEF 2011). Thus, private extension services play a crucial role in agrarian change process in Kyrgystan and Tajikistan, since not much is being done by the state to support agricultural producers.

Differently from other former Soviet republics input supply in Kazakhstan is arranged relatively well, especially thanks to financial strong state support. In addition, there are state subsidies to purchase inputs for farmers. In Kazakhstan extension has shown to play a bridging role, bringing together the input providers and farmers (GIZ/ZEF 2011, interview with Yusupbaev J., director of the centre of knowledge dissemination 'Tassay', 2010). In particular, extension offices offer pricelists of input suppliers and offer them to farmers. And the other way around, if the farmer needs something, he enquires to the extension office or via the hot line. Extension officers will offer them what is there on the requested good. The representatives of the seed and fertilizers' companies can come to the seminars and advertise their products. As a perspective to develop, now Kazakh extension intends to combine state and private activities, meaning to start additional services with additional payment (Centre of knowledge dissemination 'Tassay' 2009).

This small overview of service supply in a few post-Soviet countries suggests that the following baseline of all former Soviet states. AGSOs evolve based on the previously available Soviet infrastructure developed for *kolkhozes* and *sovkhozes*, rather than starting from scratch (Goldberg et al. 2011). Provision of services in ex-Soviet states can be split into purely state supplied, where the state has significant financial basis presented by the example of Kazakhstan.

Service supply can be organised via the private sector, state or via NGOs as presented by the case of Ukraine. In Tajikistan, Kyrgystan, Georgia, and Azerbaijan service supply is heavily supported by international donor organisations (GIZ/ZEF 2011; interview with Isgandarov Y., programme coordinator of Agro Information Centre, Azerbaijan). Another similarity amongst these countries is that there is no state control and no significant state financial allocations to the development of AGSOs (except in Kazakhstan). Moreover, an important role is performed by NGOs and extension services who try to streamline donors' money into the direction that would fit best to the agricultural producers of each country. However, the amount of financial allocation to development projects in agriculture in these countries by donors is very low. In addition, a private sector of input supply has emerged in those countries importing fertilisers and other inputs from abroad.

Uzbekistan differs from the other post-Soviet states, as service provision mainly serves the state procurement system in agriculture and is state driven in contrast to all presented countries. The input supply in the selected ex-Soviet states was run by the private sector, international donor organisations and little state support. Agrarian change and the role of the state in other ex-Soviet countries differ from Uzbekistan. Therefore, it is difficult to adapt the experience from other countries with regard to restructuring agricultural service organisations in the Uzbek agricultural model.

2.6 Rationales and behaviour of agricultural actors in Khorezm

The actors discussed in the thesis, in particular the Uzbek government (or the state[7]), bureaucrats and local elites, farmers, and agricultural service providers have ideological and economic or material interests. *The ideological dimension* includes culture, traditions, and norms. *The economic or material* dimension includes what actors do to achieve their interests while functioning in the present economy. To which strategies they appeal to maximise returns from their economic

7 State and government are used interchangeably in this thesis. I draw on definitions of 'state' and 'government', as defined by Hyden et al. State refers to all institutions (government organizations) that make the public sector with a responsibility for implementing policy. Government refers to elected or appointed officials serving in core institutions at the national, provincial, country, city, and local levels. Therefore, while referring to the 'state' in the thesis, I mean any government organization acting in the name of the state strategic interest in cotton and wheat.

activities? As it will be shown, all actions of the above mentioned actors are guided by these two interests.

***State interest*:** The Uzbek nation state has multiple economic interests in the agricultural production of Uzbekistan. An analysis of the reforms shows that the state is trying to find the optimal way to organise agriculture that would secure the employment of the rural population, national food security, as well as the returns from cotton production, Uzbekistan's leading export product (Kandiyoti 2003; Khan 2005; Spoor 2006). Since the cotton sector is very important for the state, the Uzbek government has decided to maintain the cotton sector by relying on actors and structures that have been available in Uzbekistan since Soviet times. In particular AGSOs, state bureaucrats, agricultural service providers and agricultural producers are mobilised in order to serve state procurement goals. The process of mobilisation of the state bureaucracy and other organisations, which were established during the Soviet times for the purposes of the state procurement system, is described in the international literature (see Chapter 4 for more details). 'Bureaucracy' denotes the hierarchical apparatus in control of all social and economic affairs and includes not only government officials and managers, but functionaries of the party and the mass organisations as well (Kornai 2008; Pham 2004). Also noticed by Kornai, the vacuum left by eliminating administrative commands, and thus direct bureaucratic coordination, is filled not by the market, but by other, indirect tools of bureaucratic coordination (2008).

***AGSOs' interests*:** The functioning of AGSOs is strongly framed by the cotton interest of the state. AGSOs have to fulfil state interests and assignments with regards to the state procurement system. Moreover, since the majority of AGSOs are rooted in Soviet times, members of the organisations continue to function in the way they used to during Soviet times. Despite strong ideological interest of the state, AGSOs have their own economic goals. Each organisational leader tries to develop its organisation and improve finances to sustain through the transition. Thus, organisations serve as the base for its employees to survive the transition period and wait 'for better times'. At the same time, organisations have to fulfil a variety of other goals set by the state.

AGSOs, supposedly, are economic actors interested in maximum profit from their activities. In reality, AGSOs are multi-faceted organisations, not only aiming to making a profit (despite their official job-description) but forced to keep an eye on many, partly competing goals. Foremost, the organisations have to support the state administration in all activities related to the production of state-ordered crops. These tensions between the goals described were observed at the regional level, but also at the district and local level.

Since AGSOs were established in the socialist system, it is challenging for them to function under the changing environment together with the state procurement system. In this complicated situation, AGSOs survive thanks to the *agency of their leaders*. Also, in many cases where there is not enough money, they find solutions to problems by appealing to *informal rules*.

The former *kolkhoz* leaders, who are the present day leaders of AGSOs, share many similar strategies to adapt to the changing environment because their education and ideology were similar in many of those states. The research by Luong (2002), Yalcin and Mollinga (2007), Trevisani (2009), Shtaltovna et al. (2011) on Uzbekistan and other former Soviet countries8 demonstrated the agency of former *kolkhoz* leaders as one of the main factors that helped them to carry on through the process of transition. The *agency of actors* includes the following features: the ability to cope with and adapt to the changing environment, protect the properties of former *kolkhozes* from the exigencies of reforms, corrupt authorities, arbitrary judgements by the tax inspection and the deformed institutional framework in general (Ledeneva 2006). In the transition process, Toleubayev (2009), Kitching (1998), Wall (2006), Oberkircher and Hornidge (2011) underline the importance of managers' leadership in preserving farms from collapse. Former *kolkhoz* leaders managed to preserve and maintain the Soviet time production infrastructure, to mobilise farm workers to guard these infrastructures from robbery day and night, to employ farm workers and support the social infrastructure of the village throughout the hardship of the transition period. Possession of a professional farming background is one of the major prerequisites to run a farming business. They also showed the ability to use one's ties established in Soviet time, their personal ties with the local authorities and agri-businesses. As suggested by van Assche et al., survival depends on political and economic support, and both usually materialize via informal channels (forthcoming). Spoor and Visser (2011) illustrated that large farm enterprises in Russia developed survival strategies and responded in rational ways to neoliberal reforms by building new social-economic networks and cultivating old ones and developing new forms of integration in an insecure

8 On Kazakhstan by (Toleubayev 2009; Spoor 2012); on Hungary by (Lampland 2002; Hann 2003); on Moldova and Russia by Spoor 'Agrarian reform and transition: What can we learn from 'the East'? Inaugural Lecture, pp. 1-33; on Russia by (Ledeneva 2006); on Ukraine by (Shtaltovna 2007).

market environment. The AGSOs studied in Khorezm confirm many of the elements of former *kolkhoz* leaders that are crucial for organisational survival and reinvention during the agrarian change period (see Chapter 7). Despite qualities of the leaders inherited from the Soviet system, farmers and other actors presently involved in agriculture lack the skills of how to work in a changing environment. As noticed by Toleubayev (2009), in the post-Soviet hostile neoliberal market environment newly emerged farmers had no time to learn and had to learn via trial and error from their own mistakes, which for many jeopardised the ability to stay in the farming business. Based on the empirical data, I will identify those skills necessary but not always available for surviving the ongoing period of transition.

The analysis of clientelist politics in agrarian transformation processes as currently undertaken by Uzbekistan therefore gains importance. As will be shown, patron-client relationships take place between farmers and state neglected AGSOs while providing services under the state procurement system (see Chapter 6); they also occur between AGSOs and the state elites and bureaucrats (see Chapter 5). *Clientelism or patron-client relationships* are identified by Clapham et al. (1982) as a model of social exchange and a specific strategy of political mobilisation and control, whilst according to Oi it is a type of elite-mass linkage through which the state and the party exercise control at the local level, and through which individuals participate in the political system (1985). An ordinary citizen, willingly or not, plays the game of private connections and unwritten rules (Mendras 2002a). They particularly emerge with two underlying processes, namely state centralisation and market expansion. Similarly as in post-Soviet Russia, as argued by Ledeneva (2006), contacts in post Soviet Uzbekistan play an important role in establishing personal relationships with patrons, partners, and clients. Similar behaviour is appreciated in the Chinese socialist context, where one of the most important benefits for a team leader is respect and the support that it entails. The client is characteristically the team leader's most enthusiastic supporter and helper, and can be counted on to praise the patron's leadership and to encourage others to do likewise (Oi 1985).

Bureaucrats' and local elites' interests: the bureaucratic system has been in place since Soviet times. If before it was well financed, now bureaucracy is underpaid. In order to survive, state elites and bureaucrats need to meet their economic interests and goals. So, they are trying to maximise their profit by relying on their remaining power from Soviet times. Local elites and bureaucrats are mobilised by the ideological state domain to meet state economic strategic interest. So, while fulfilling tasks assigned by the central government, local elites and bureaucrats appeal to patron client relationships, corruption or rent-seeking.

With regards to the role of local elites and bureaucrats in the state procurement system and their survival strategies during transition, the concept of *patron-client-relationships* illustrates the political behaviour of low-status actors, as they are incorporated, recruited, mobilised or inducted into the national political process (Powell 1970). In addition to these traditional patrons transformed into brokers, other local people with 'outside connections' also can assume brokerage functions – schoolteachers, physicians, pharmacists, priests, tax collectors and other local officials (Powell 1970). These people are referred to as 'small intellectuals' of society, whose status and role functions place them in the 'strategic middle' of the social structure (Paulson 1967). In the Uzbek context, organisations such as the tax office, the public prosecutor's office, the police and state technical supervision departments play the role of 'state brokers'. Apart from their direct responsibilities, they have to make sure, that AGSOs and farmers follow state orders. Markowitz (2008) in his work has pointed to the '*prokuratura*' (public prosecution) as informally one of the most powerful offices within Uzbekistan's state apparatus. Using the example of Samarkand Province, Markowitz illustrates how an attempt to extend state power to the regional level inadvertently strengthened local elites' control over extractive processes and entrenched patterns of 'stationary' *rent seeking*. The use of patronage to rule regions, however, has promoted different patterns of rent seeking within the country, which in some localities further entrench the authority of local elites while in other localities facilitate predation by state officials. As local patterns of rent seeking multiply and become embedded within the state's territorial infrastructure, their diversity may pose a significant challenge to the central state's command over resources in the future. According to Luong (2002), instead of motivating elites to reject their previous political identities, the transition signalled the necessity and desirability of reinvesting in the regionally based patronage networks that undergirded their political power. In chapter 5, I provide empirical examples on the bureaucracy's ability to adjust their interests to the changing environment of Uzbekistan's nation state.

***Farmers' interests*:** farmers, similarly to AGSOs, are bound to the state procurement system. 80% of land they "own" is used for state procurement crops. Despite the strong state economic interest imposed through the state ideological interest, farmers are entrepreneurs and try to find ways how to find additional income from the state procurement system. One main method is through the production and sale of other commercial crops. In order to secure a source of livelihood for the population without significant financial support from the government, there is the domestic peasant economy with its relative independence from the centre lying parallel to the cotton economy (Ilkhamov 2000).

Similar to AGSOs, farmers, when financial resources are slim, turn to *informal ways* of arranging things and to *kinship relations* (for more details see Chapter 6).

Working under the state procurement system in a period of agrarian change involves many *informal arrangements* between AGSOs and government organisations, as well as between AGSOs and farmers. The service providers studied have to function in a governance context marked by a discrepancy between formal and informal institutions (Ostrom 1990; North 1998), whereby formal institutions gradually acquire the character of a façade with doors and windows that are used every now and then, selectively. As argued by Ledeneva, informal practices are an integral part of the post-socialist transformation (2006). In order to work in the transition environment, these are the *'rules of the game'* that matter most because they determine who will set future 'rules of the game'. Thus, they determine not only who will govern, but also the manner in which they will govern (Luong 2002). The texture of informal social relations between leaders and members (in this case local elites, bureaucrats and AGSO) can be very revealing, as hierarchies are reproduced through ordinary daily activities (Fox 2007). As will be presented in chapter 4, being part of the state administration does not give as many advantages as in the past. However, the previous power and authority held by the state from Soviet times can be mobilised to assist in implementing a great range of different survival strategies of state bureaucrats during the transition process. The 'rules of transition' are short-term, as the situation is changing at such a rapid pace. Consequently, those who are in power try to make maximum use of their positions and thus abuse farmers and AGSOs as the following case studies will illustrate. Within this context, actors are more likely to adapt strategies oriented towards short-term rather than long-term distributional gains because the potential rate of change and high degree of uncertainty limit their ability to 'predict' far in advance (Hornidge et al. 2011; Van Assche et al. forthcoming). This is also in line with Luong's assessment of local bureaucrats, making use of their positions in the state apparatus (Luong 2002). Local elites pursue political liberalization only when it is either consistent with their beliefs or necessary to ensure their political survival (Luong 2002).

Based on these preliminary rationales of each actor, the following theoretical predictions of each collective actor's behaviour in the current transition period can be suggested.

State: is likely to continue and possibly increase control over agricultural producers in order to meet its strategic economic goals.

Bureaucrats: in order to survive, state elites and bureaucrats continue to appeal to rent seeking and corruption by means of AGSOs and farmers. From another perspective, if the bureaucrats received better salaries, possibilities to fight corruption would increase. The close interaction between AGSOs and state elites is rooted in the Soviet planned agriculture, and also it is related to a broader *centre-periphery* discussion, which gained a foothold in pre-Soviet and Soviet Uzbekistan (Ilkhamov 2000). There is a historical divide and differences between the Khorezmian and central government actors. The antagonistic relationship between centre and periphery produces what Ilkhamov has described as the "battle for cotton" (2000). After the end of the Soviet Union, the Uzbek President Islam Karimov's politics have weakened the provincial governments and concentrated power in his hands (Luong 2002). In addition to this, the overall Soviet maintenance system has ceased (see Chapter 3). These two conditions have put the still existing regional/local state apparatus into the position where they have to seek for ways to carry on without state support. In chapter 5, I empirically demonstrate how AGSOs play a role in the state organisations' survival strategies. The cotton revenues mainly go to the top and leave little on the provincial and local level. This situation reinforces the centre-periphery divide. At the same time, detailed information on the agricultural production processes and the challenges involved in keeping them going experienced by farmers and AGSOs are not fully communicated to the top, resulting in little awareness and thus little assistance from the center. Provincial and local level elites consequently play their own game. If not receiving revenues from cotton, they secure their pockets by means of farmers and AGSOs by using a wide range of formal and informal means of rent-seeking. AGSOs, like farmers, are caught in a web of conflict between the center (i.e. the state) and the periphery, whereas the fight for returns from cotton is negotiated between Tashkent and provincial and local elites. The Uzbek state consequently has a twofold approach to present day agriculture. On the one hand, the central government puts all-round pressure on local elites carrying out selective but periodical purges of local staff with the purpose to prevent corruption from becoming too rampant and resources from being redistributed in favour of provincial elites. On the other hand, it still stops short of completely undermining the economic basis of the local elite's power, thus securing the loyalty and obedience of the latter in return (Ilkhamov 2000). Looking at the interactions between AGSOs and state elites as well as at the inner organisational dynamics of AGSOs, chapter 5 focuses on AGSOs as a space for local elites' and bureaucrats' personal agendas in servicing their livelihood strategies. This contributes to the analysis of agrarian change processes in Central Asia during transition.

AGSOs: The majority of AGSOs are strongly embedded in the state procurement system. As it will be illustrated, some AGSOs are looking for diversified sources of income. They try to allocate more time to provide services to clients outside the state procurement system to increase their income. Other AGSOs feel comfortable fully serving the state procurement system.

Farmers: As it will be shown, farmers have a strong economic interest in growing commercial crops and in this regard they manage to mobilise the monetary means to obtain inputs. Since there are difficulties involved in dealing with the state procurement system, like delays with payments and the state allocated subsidies not being enough to cover all production costs, farmers' motivation continuously declines towards growing state ordered crops. For such crops, they try to reduce the costs for inputs by, for example, appealing to kinship and informal arrangements to either pay less or receive them faster. Similarly, farmers purchase patrol-driven pumps and repair old diesel pumps in order to be able to irriage their kitchen gardens when the electricity is cut off (Oberkircher 2011).

Assessing these rationales and theories in combination indicates the range of different types of service providers emerging as part of the ongoing transition period. In the next section, I will present the methodology applied and a relational typology of AGSOs that I have developed as a heuristic tool to study the differences between the multiple types of AGSOs.

2.7 Methodology

> 'Aggregate statistics and compendia of decrees and laws tell us little without complementary close descriptions of how people – ranging from farmers to factory workers, from traders to bureaucrats, from managers to welfare clients – are responding to the uncertainties they face'.
>
> (Burawoy and Verdery 1999)

Data was collected through various qualitative methods and a multi-stage approach between 2009 and 2010 for a total of 11 months. The research workplan was as follows: mapping of agricultural service providers, work with secondary data, i.e. looking at the relevant laws and administrative regulations, conducting farmer' survey, conducting internships in three agricultural service organisations, interviews with decision makers and experts in agriculture on local, regional and national levels, interviews with all agricultural service providers available in Khorezm region, and interviews with actors involved in extension services in other ex-Soviet countries.

Farmers' Survey

Quantitative and qualitative data were collected through the farmers' survey on service provision by agricultural service organisations (see Annex 1 for a questionnaire for farmers and Annex 2 for the farmers' responses). The survey was conducted amongst farmers from five Water User Associations (WUAs), which are located in five districts that differ in biophysical, social and institutional conditions (see Figure 2). Abdullaev et al. (2008) distinguished the following indicators for the choice of the Water User Associations' locations:

1. Remoteness from the water source;

2. Relative water scarcity;

3. Social situation, living standards, diversity of agricultural activities;

4. Institutional strength and type of water management.

Selecting farmers according to the above mentioned criteria was useful as it allowed me to capture the diversity amongst different types of farmers and their activities. The survey was conducted with 50 farmers growing different types of crops. The farmers' survey was conducted in the summer of 2009, which was the first agricultural season after farm consolidation (see Chapter 3 for details) at the end of 2008.

The respondents were asked to reflect on their needs regarding service provision, and to identify the services that were the most important to them. Also, the respondents were given questions such as: what kinds of services were in demand and were delivered; what kinds of services were delivered and were not in demand; what kinds of services were in demand but were not delivered. Based on the results of the survey, I inferred what kind of services they needed. I focused on identifying the reasons why some particular services were the most important and the problems related to them. In addition, I examined the services that were lacking and the steps undertaken by farmers to obtain such services. Alternative possibilities of service provision were investigated in the province of Khorezm.

Farmers for the survey were selected according to the categories of agricultural forms of production in the Khorezm region by Veldwisch (2008). The categories are: (1) state-ordered production, which includes the production of cotton and wheat; (2) commercial production, which involves growing of rice, horticulture, poultry production and to a lesser extent the production of vegetables and fodder, and also dealing with animal husbandry; (3) household (subsistence) production (*dekhan* farms). *Dekhans* grow fruits and vegetables in their gardens, and wheat and rice on their small plots of land. This kind of production primarily aims at home consumption and includes barter arrangements as well as petty trade at local markets.

These production types differ according to the size of the allocated land, labour involved in production process, ownership, land tenure and production specialization among others. These three forms of farming are strongly interconnected. In practice they overlap, as the same household can participate in different forms of production. Nevertheless, each form of production has its own rationale, socio-political control system and economic characteristics. Each type of production has different needs and demands. The farmers' survey serves as a basis to capture which services are available and farmers' opinions in regard to the services provided.

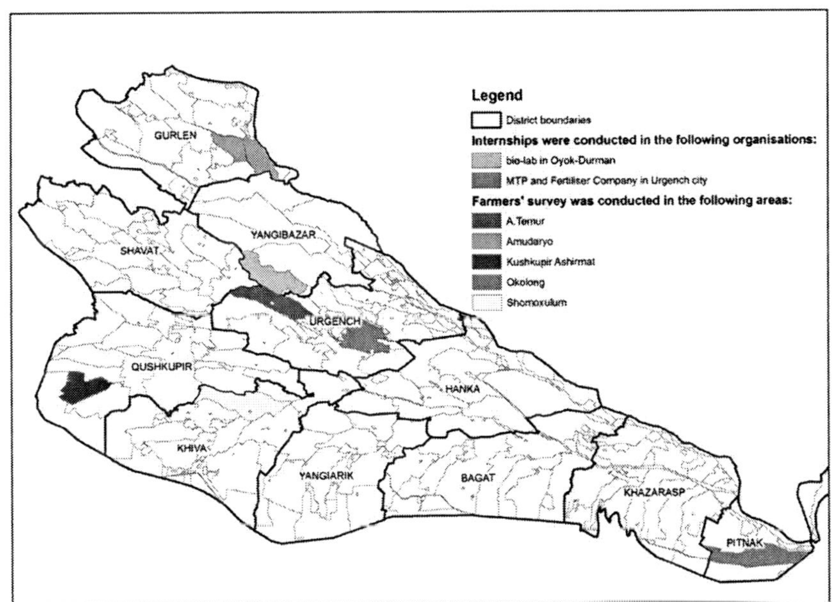

Figure 2: Map of the Khorezm province indicating where farmers' survey and internships took place. Source: The map prepared by A. Lee, GIS laboratory, ZEF/UNESCO project, Urgench 2011

Literature review

A literature review was carried out focusing on legal and scientific publications on agricultural service provision in Central Asia (See Chapter 4).

Interviews

30 semi-structured expert interviews were conducted with local, regional and national government officials. Interviews were also conducted with all available agricultural service providers (including private ones) at different management levels in Khorezm province (see Annex 3 for a list of interviews conducted). 85 semi-structured interviews were conducted with the AGSOs' staff as well as 15 expert interviews with additional key informants. Furthermore, 20 meetings of the AGSO's directors with state representatives as well as trips to farmers' fields were attended.

Interviews with Extension service providers in other ex-Soviet countries

I conducted interviews with the representatives of NGOs, state organisations, universities, international development aid in regard to extension service systems in other ex-Soviet countries: Tajikistan, Kyrgyzstan, Uzbekistan, Kazakhstan, Ukraine, Georgia, and Azerbaijan (see Section 2.4)[9]. The data obtained through these interviews helped me in the framing of my research and further analysis of my research findings.

Case Studies

To analyse the external and internal environment of AGSOs, I decided to use a case study approach by means of internships and interviews with organisations providing three different services /input: machinery, fertiliser, and pest-control. I interned at the following three organisations providing these services to the procurement system: a Machine Tractor Park (MTP), a Fertiliser Company (F. Co), and a private Bio-laboratory (Bio-lab). In addition, I conducted interviews with employees of other organisations providing or related to the three services of interest, such as alternative Machine Tractor Parks and state bio-laboratories, for data triangulation purposes. Interviews were also conducted with employees at Ifoda, a private organisation established recently with no ties to the state.

All fieldwork interviews and communication took place in Russian and Uzbek, in most cases without a translator. The names of some respondents will not be given and some names will be fictitious for reasons of privacy.

---

9 In 2010, I attended a GTZ – ZEF/UNESCO Symposium on agricultural service provision in newly independent states of the former Soviet Union conducted in Urgench, Uzbekistan. There I met a variety of individuals involved in agricultural service provision in ex-Soviet countries.

The organisations were selected based on the results of the farmers' survey and initial interviews conducted with different providers. I originally intended to select the services most mentioned in the survey (either those lacking or most important to farmers). However, the most mentioned service, water supply via WUAs, has been extensively researched by the ZEF/UNESCO project and was not a suitable choice for further research. The next most frequently mentioned service was fuel supply. Unfortunately, there were no internship opportunities available with a provider of this service. Therefore, I decided to focus on the third and fourth most frequently mentioned services to the farmers surveyed: machinery services and fertiliser supply.

While conducting the internship in the MTP and initial interviews with other service providers, it became obvious that not all agricultural service providers were doing equally well. They received different levels of support by the state administration, and some of them were more state-centred then others. One service that was developing poorly and was not identified as necessary to farmers was bio-laboratories. Therefore, I decided to also investigate bio-laboratories as a comparative approach to the other two providers and interned at one.

Internships

'All social systems, no matter how grand or far-flung, both express and are expressed in the routines of daily social life...' (Giddens 1984).

The main purpose of the internship was to understand how the organisation functions in the period of transition. As such, the internships were used to study the organisation as an insider in order to examine how agrarian change processes have impacted agricultural service organisations. Analysis of day-to-day activities of AGSOs allowed me "to extract the general from the unique, to move from 'micro' to 'macro', to connect the present to the past in anticipation of the future, all by building on pre-existing theory" (Burawoy, 2009: 21). I approached my study through 'organisational ethnography' as described by Schwarzman (1993). By ethnography I mean writing about the world from the standpoint of participant observation (Burawoy 2009). By conducting ethnography within organisations, ethnographers learn about the cultures and structures of organisations from the inside out and examine both what people say and what people do (Schwarzman 1993). Organisational ethnography is crucial because as a research method, it problematises the ways that individuals and groups constitute organisations (and societies) on a daily interactional basis (Schwarzman 1993). An ethnographic approach involves documenting key actors' actual practices, exploring both how they understand their own goals and their external environments (Fox 2007).

The first internship was at a Machine Tractor Park, the second at a F. Co and the third at a private bio-lab 'Aek Durman'. I spent from three to nine weeks in each of the organisations. Apart from legal documents and financial reports, qualitative data in large were collected by means of daily participant observation of the processes in the organisation, semi-structured interviews with the staff, participation in daily activities of organisation, and attending internal and external meetings. I followed individual kin members, specific groups (insiders and outsiders) and observed their routines of work. I also participated in events such as birthday parties, trips to farmers' fields and meetings with state representatives dedicated to the state agricultural campaign. This empirical study is thus to a great extent based on the direct observation of the behaviour of the members of the three service providers.

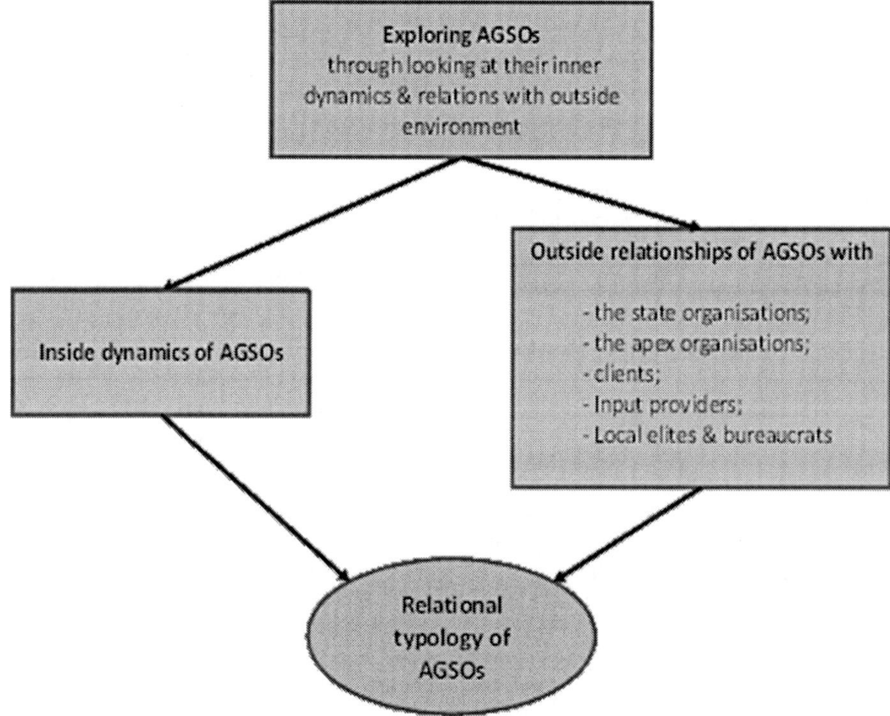

Figure 3: Relational typology of AGSO. Source: author's compilation, 2010

Case analysis: Relational typology

While doing research on organisations, I found that a lot of studies focused on the inside relations and interpersonal relations in the organisation (Schwarzman 1993). Examing the outer relations of AGSOs opens a window to agrarian change process, the interactions include relations with the variety of state organisations, clients and input providers. As Burawoy commented on the study of Warner and Low on Hawthorne in 1947, 'apart from looking just at human interactions, moving beyond considering only shop floor conditions to an examination of 'the economic, political, and social forces which were transforming relations between capital and labour in the 1930s' (1979). So we can identify and analyse the mechanisms that link structures and events, as well as understand how organizations manage to sustain themselves through the agrarian change. Thus, in order to validate the typology of AGSOs in Uzbekistan, I will look at relations of the three studied organizations with their inner and outer organizational environments (as illustrated on the Figure 3). According to Chapple, organisations can best be studied as a system of relationships between individuals (1941: 6 in Schwarzman).

To understand the complex determinants of organisational change, I developed an analytical tool that hypothesized different types of AGSOs. The relational typology tool suggests that there are three different types of AGSOs during transition process. The first group consists of *state affiliated agricultural service providers*. Organisations of this type are of strong economic interest to the state and under state vigilance. The state affiliated agricultural service providers fulfil the state orders as required by the present-day procurement system. Therefore, they get more economic state indulgences but are also more bureaucratic and overstaffed than the other two types of service organisations. The next category is the *state ignored agricultural service providers*. They are of little economic interest to the state since service provision by this type of organisation is not required by farmers who cultivate cotton and wheat. Moreover, state ignored agricultural service providers receive less state-support than state affiliated service providers and are least covered by the monitoring mechanisms applied to the service organisations involved in producing state-ordered-crops. State ignored agricultural service providers operate both for the state procurement system and outside of it. The producers that are not subjected to the state procurement system (e.g. commercial farmers, animal husbandry farmers and horticulture farmers) are the primary users and customers of services offered by state ignored agricultural service providers. The last category consists of *state neglected agricultural service providers*. They are in the process of transition from being state affiliated service providers to becoming state neglected service providers. This type of organisation remains vital for all agricultural production systems. Due to the transition process, state neglected service organisations are withering away since the state has (gradually) reduced its support, forcing them to enter a process of re-structuring. The staff of state neglected service providers still knows the way of running agriculture as it was practiced during the Soviet times and function as a taskforce to manage present agriculture – which is of strategic importance to the state.

As a framework to analyse the relationships of AGSOs, I created categories for the observations and data I gathered during my internships (Figure 4). Analyzing empirical data in the suggested order helped to identify similarities and differences between the three organisations and contribute to a relational typology of AGSOs.

Based on my research findings on the dynamics and relationships of the three types of organizations, similarities and differences between them were identified and used to validate or modify the suggested organizational typology.

Data analysis

The data from the internships, interviews and farmers' survey were analysed using Atlas.ti software[10].

10 For more details on Atlas.ti visit www.atlasti.de

Organisations	Alignment while keeping distance: the negotiation of relations between agri cultural service organisations and the state	Personal agendas and organisational means: agricultural service organisations as platforms of entrepreneurship for employees and state elites	Servicing decollectivisation: state order versus private commercial production clientelism	Reinvention and diversification of agricultural service organisations: The importance of leadership
The Fertiliser Company				
The Machine Tractor Park				
The Biolaboratory				

Figure 4: A concept for case analysis. Source: author's compilation, 2010

CHAPTER 3: AGRARIAN CHANGE AND RURAL TRANSFORMATION

3.1 Introduction

In this chapter, I analyse in detail agrarian reform and sector restructuring in relation to AGSOs. I address the question what kinds of agricultural service providers have emerged in the twenty years of the 'reform'? I do so by looking at how AGSOs have evolved through the Soviet period, and how AGSOs have evolved to match the changes in the agricultural sector within the past twenty years of independence in Uzbekistan. I will also look at farmers' feedback towards services provided by AGSOs. This chapter lays a foundation for the following empirical chapters.

Uzbekistan has gone through sixty years of Soviet collectivised and planned agriculture and twenty years of post-Soviet agriculture, which included three major reforms where aspects of the Soviet and post-Soviet systems became intermixed. The Soviet history and pre-Soviet period as well as recent globalisation processes shape Uzbekistan's rural social, political and economical landscape (Ilkhamov 2000). AGSOs were established as a part of the Soviet collectivised agriculture to serve the needs of the collective and state farms. AGSOs have been defined as the network of service and input providers for agricultural producers (Niyazmetov 2012). 2004 was a turning point for AGSOs' evolution when their clients changed drastically from a small number of large, powerful and autonomous *sovkhozes* and *kolkhozes*, to a large number of small, vulnerable, under-financed farms. In the twenty years of independence, AGSOs have been forced to reinvent themselves to sustain themselves in this transitional period. A spectrum of outcomes in terms of agrarian structure, organisational changes of AGSOs, state influence and changing relations between state and agricultural producers have occurred. In the current stage of the agrarian reform AGSOs are located at the interface of the state and farmers in contrast to the past, when AGSOs were a part of the state agricultural mechanism. However, as will be obvious, AGSOs are still levers of control for the state in present day agriculture. The situation in agriculture is still in flux and undergoing changes.

An overview of agriculture during the Soviet period is presented in section 3.2 followed by a description of post-Soviet agricultural reforms in Uzbekistan in section 3.3. The present day agricultural situation presented in section 3.4

demonstrates the daily challenges of AGSOs and farmers under the present agricultural frame. Section 3.5 will present a detailed overview of AGSOs with their shortcomings reflected by farmers. The case studies of the Fertiliser Company, a Machine-Tractor Park and two pest control organisations in Khorezm, Uzbekistan are presented in section 3.7. The chapter closes with conclusions in section 3.8. The account is primarily based on fieldwork data since there is hardly any literature available on this issue.

3.2 Agriculture and service provision during the Soviet period

As suggested in section 2.6 there is a path dependency in agriculture built up over decades of central planning during Soviet times. Here I will draw attention to agrarian structure of path dependency (Thelen 1999; Sydow 2005). In particular, I look at how the previous Soviet agrarian structure of *sovkhozes* and *kolkhozes* shapes the social relations of production in present day agriculture. The *kolkhozes* and the *sovkhozes* were the central productive units in the Uzbek rural sector, virtually comprising all arable land. For all practical purposes they combined functions of a mega-farm and a local government (Humphrey 1998; Ioffe 2006; Trevisani 2007; Allina-Pisano 2008). In Soviet Uzbekistan, farming was predominantly cotton production requiring intensive labour employment and complex irrigation schemes (Trevisani 2008). The *kolkhoz* was a site of coordination for many sub-organisations and for policies and resources coming from outside (Van Assche et al. forthcoming). While such a utilitarian and monofunctional organisation of the countryside obviously introduced rigidities, the autonomy of the *kolkhoz* also offered possibilities for local adaptations. *Kolkhoz* management in many places functioned more or less democratically, and had the power of local governments in the western tradition. Especially where agriculture was profitable, collective farms could generate sufficient resources to plan, organize and develop according to their own standards (Kucherov 1960; Humphrey 1998; Verdery 2003; Ioffe 2006; Allina-Pisano 2008) Usually, there were production targets, but diversification was often possible outside these targets, and the targets themselves were often negotiable for active *kolkhoz* managers – if not on paper, then in practice. What later became agricultural service organisations, existed partly in and partly outside the *kolkhoz*. Spare parts for machinery and necessary inputs were provided to *kolkhozes* upon demand. The collective farms functioned often as a shelter from the direct effects of higher- level policies (Hough and Fainsod 1979; Humphrey 1998). Within the *kolkhoz*, new ideological fashions, new rhetoric, new ambitions would lose their sharp edge and could often be twisted and turned in such a way that something useful for the local community (or the local elite, depending on the case) emerged (Van Assche et al. forthcoming). Within the *kolkhoz*, party representatives formed links with the communist party hierarchy. Thus, individuals made their careers by moving from *kolkhoz* to politics and the other way around (interviews, Hann 2003; Verdery 2003; Ioffe 2006).

Apart from a productive function, *kolkhozes* and *sovkhozes* had a social function (Spoor 2012). It pervaded the life of its rural inhabitants as a 'total social institution' that encompassed the whole range of their political, cultural, and economic relations (Trevisani 2008). The Soviet policy made the Soviet farm an economic production unit and a socio-political institution (Toleubayev 2009). Whilst in most Western economies, these social security functions (education, health, and pensions) would be in the hands of national and local authorities or agencies, in the former Soviet Union these were quite often channelled through collective and state farms. The most basic human needs, which are security and protection, were almost satisfied. Social needs were met at a fairly high level, however, services were not carried out under the initiative and wishes of workers, but organized from the top. *Komsomol*[11], party and trade union organisation on the ground conducted various meetings, which were attended by all members of the workforce. On the agenda were various issues such as socio-political matters. Such activities simulated worker participation in company management and helped to increase workers' self-esteem. Active participants in the labour process and social activities were noticed by the leaders and moral encouragements were given, i.e. certificates of honour and appreciation, marks of distinction, and were awarded with orders and medals, one's name would be used as an example to others, a trip to a resort or a financial remuneration was granted. Photos of 'Shock workers of Communist labour' (*Udarnik*) were hung on the leader board, so that workers knew their hero. Additionally, a financial premium was allocated, the minimum size of which was minimum ten rubles (Encyclopedia of Management 2011).

The preceding socialist system laid a foundation for present day agriculture. It included basic functions of controlling social activities vital for survival, it organised production and supply to the public of the goods and services needed to survive, it insured in its own way the discipline required for the coordination of activities and for human coexistence, and it founded a legal and moral system in which people could find their bearings (Kornai 2008).

---

11 Komsomol (from Russian) - was the youth division of the Communist Party of the Soviet Union.

3.3 A chain of post-Soviet agrarian reforms: 1991 – 2011

The breakup of the system of collective farms differed in every post-Soviet country. Uzbekistan followed a very different agrarian transformation pathway from any of the Eastern European states. Very gradual, state-led reform took place in the largely agricultural republic (Spoor; 2010: 16). Uzbekistan chose to be very careful with privatization of land and agricultural production and designed its own path of reform (Van Assche et al. forthcoming). As illustrated in figure 5, between 1991 and 1998, the agricultural system continued to function in broadly the same way as it used to during the Soviet times, that is, production activities remained organised in the form of *sovkhozes* and *kolkhozes* that were slowly transformed into joint stock companies (*shirkats*) (Khan 2005). The *shirkats* were kept within the planned system, which remained largely in force, with planned production quota, delivered inputs, and procurement of output, in particular cotton. This implied a largely cosmetic change in the form of production (Khan 2005; Spoor 2012). Cotton was (and is) a 'king' in Uzbekistan and the most important foreign exchange earner of the economy (Spoor 2010).

In the late 1990s, a second phase of agrarian reform was initiated, focusing in particular on the financial sanitation of the (many) insolvent *shirkats*. Until then, there had been some reforms within the labour regimes of the *shirkats* (Spoor 2004). Labour reform in the *shirkat*, on the one hand, gave them more independence, but on the other hand, control was retained by the government with continued direct government funding of agriculture (Ilkhamov 2000).

Till the mid-2000s, more substantial changes took place in the agricultural sector. In this period, *shirkats* were 'privatised' and subdivided into small, individual farms (Veldwisch 2007; Lerman 2008; Trevisani 2008). The agricultural sector in Uzbekistan was organized into a system of farms growing cotton and wheat, in combination with the existence of small family plots for household food production. 'Private' farms were established, but without private ownership of the land and still tied to production targets for cotton and wheat (Trevisani 2007; Veldwisch 2008; Djanibekov et al. 2010). In 2004 the number had grown to 103,900 covering an area of 2.9 million hectares, while by early 2008 the number had more than doubled to 235,000, encompassing 5.8 million hectares of land (Dept. of Statistics of Ministry of Macroeconomics and Statistics, Tashkent in Trevisani 2008). Uzbekistan has fully transformed the large *shirkats* into medium-sized 'private' farms, in addition to a mass of peasant farms and household plots. Thus, the bulk of the agricultural production of cotton and wheat was subsequently carried out by farmers who hardly had a choice in what they want to produce. They are embedded in the still existing procurement system in agriculture (Trevisani 2007; Trevisani 2009). This process of de-collectivising land crucially modified social relationships within the agricultural production system, as well as between the now diverse group of agricultural actors and the state (Trevisani 2008).

Despite legislation to encourage changes in land tenure and farm organisation, all land remained property of the state (Lerman 2008). In November-December 2008 (within less than a month) farmland under the cotton and wheat state plan was reconsolidated again, merging several individual farm enterprises (of 10-25ha each) into bigger farms (of 75-150ha). The selection of farmers who continued to remain farmers or became landless depended on their performance with regard to the production of state-ordered crops in the previous years, according to official records (Djanibekov et al. 2010). Similar adjustments, but to a smaller degree, were made once more at the end of 2009. This reconsolidation of farm land (the local term is 'farm optimization') reduced the number of farms, simplifying water management and the procurement process (Djanibekov et al. 2010). After this process was completed, the cotton and wheat farmers subjected to the state procurement system usually have land holdings over 100 ha. They are called '*fermers*' and deal intensively with the service organisation. The same farmers grow free commercial crops. In addition, there are peasant farmers called "*dekhans*" with approx. 0.25 ha of vegetables or fruits, for subsistence and independent from the state plan (Veldwisch 2008; Trevisani 2009; Van Assche et al. forthcoming).

Apart from keeping cotton and grain production under control (see Chapter 4), the central government also sets artificially low prices for the government's purchase of cotton and then sells it abroad at world market prices (Luong 2002). This negatively affects the financial viability of farms and provides logic for scale enlargement. Superficially, the economy looked different from the past: it looked more like a market, however, in practice, reform really involved only reorganisation of management functions. The management functions as finance, operative management and distribution of output were separated from each other and institutionalized. On the whole, both the directive management of the cotton and grain sectors and the hierarchical structure of their organisation have remained basically unchanged (Ilkhamov 2000). Stability in agriculture is based on a divided economy, on coexistence of export oriented cotton and on the domestic peasant economy that is oriented primarily towards domestic markets, and primarily the production of food. These two economies have different principles of organisation, functioning, and most importantly, distribution of resources and output (Ilkhamov 2000). Most importantly, the government realizes that the existence of domestic peasant economy with its relative independence from the Centre solves a number of problems, one of them as being the securing of sources of livelihood for the population without significant financial support from the government (Ilkhamov 2000). Thus, post-independence reform has led to a dual or divided economy, which is composed of bigger farms, growing cotton and wheat that deliver to the state in accordance with the state procurement system, and peasants with small plots of land for growing fruits and vegetables for own consumption and selling parts of it on the local small scale markets (Ilkhamov 2000; Kandiyoti 2003; Spoor 2004; Jozan 2008). The reforms that took place in Uzbekistan during the past 20 years are an example of a gradual, massively introduced and state-led reform (Spoor 2012).

The Uzbekistan transition process is a truly unique 'social laboratory' in which profound agrarian transformation is taking place (Spoor 2012). The process of de-collectivising land crucially modified social relationships within the agricultural production system, as well as between the now diverse group of agricultural actors and the state.

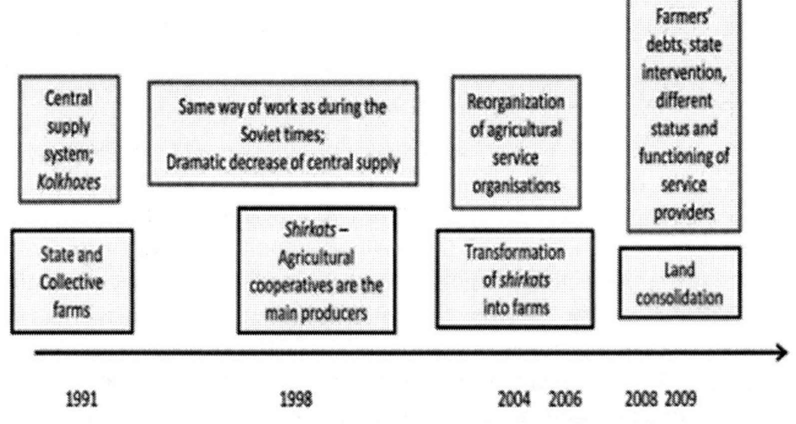

Figure 5: Reforms in agriculture and changes in service provision in agriculture.

Source: author's compilation 2010. The upper row of boxes indicates the changes in agricultural provision through the reform period. The bottom row reflects the changes in the organizational form of agricultural production.

3.4 Present-day agriculture in Khorezm: Between communism and capitalism

In this section I will analyse the practicalities of the present day agricultural system in Uzbekistan. The present system of governance, established during Soviet times, with highly centralised state power, strong vertical hierarchies and top-down rule, heavily relies on the use of state control, planning and intervention in many sectors of the economy, particularly in agriculture (World Bank 2011). Uzbekistan has largely adapted the Soviet agricultural procurement system, where farmers have to fulfil the wheat and cotton production goals assigned by the government. In particular, agriculture is managed according to five principles of the President of Uzbekistan. The five principles of the Uzbek model of development are: (1) Economy is above politics; (2) The State is the main reformer; (3) Superiority of Law and following it; (4) Running strong social policy (5) Step-wise transition to the market economy (Karimov 1995). This implies that the development path towards market economy and in particular agriculture remain to be regulated by the central government.

At the state level

For the crops grown under the state procurement system, farmers and service providers alike depend on government loans (3 percent interest rate) granted to farmers to buy inputs. How much subsidy farmers receive for growing cotton and/or wheat is decided at the Ministry level and signed by the President (see Figure 6 and Annex 6). Demand for cotton is estimated based on the norms of expenses for growing cotton and wheat and yield per ha of these crops (Interview with an official, Regional MAWR, 2009). Based on the data from the regional Ministry of Agriculture and Water Resources (MAWR), the Cabinet of Ministers of Uzbekistan estimates the plan for growing wheat and cotton. The Cabinet then orders the loan for farmers from the Ministry of Finance. The Ministry of Finance decides on the amount of the subsidised loans allocated to farmers to grow cotton and wheat and the price of which the crops will be purchased. Subsidized loans are dispensed (from the special state procurement system Fund) via the Central bank to each region by the Ministry of Finance. The money arrives at the district commercial banks and is used by farmers to pay for services and inputs needed to fulfil the state order of cotton and wheat. The commercial banks allocate money to farmers and control the use of the state subsidised loan by farmers.

The amount of fuel and fertilizers allocated to farmers is also decided on the ministry level and requires approval by the President of Uzbekistan (Interview with the officials at the department of supply of fertilisers for the state-ordered crop production in Uzbekistan, June, 2010, Tashkent; interview with officials at the state committee of de-monopolisation, support of competition and entrepreneurship, Urgench, 2009). Through the years of independence, the allocation of funds to different service providers for the purposes of the state procurement system has varied slightly (see Annex 6 illustrating allocation of money by the state for the agricultural needs to different service providers, 2008-2010). To provide necessary inputs for the state procurement system, the Uzbek industries producing fertilisers, oil and gas products are mobilised. They are given a plan to produce fertilisers or to provide fuel for state agricultural purposes. The supply of these inputs is controlled from the national to local levels through the corresponding agricultural service providers at the local level.

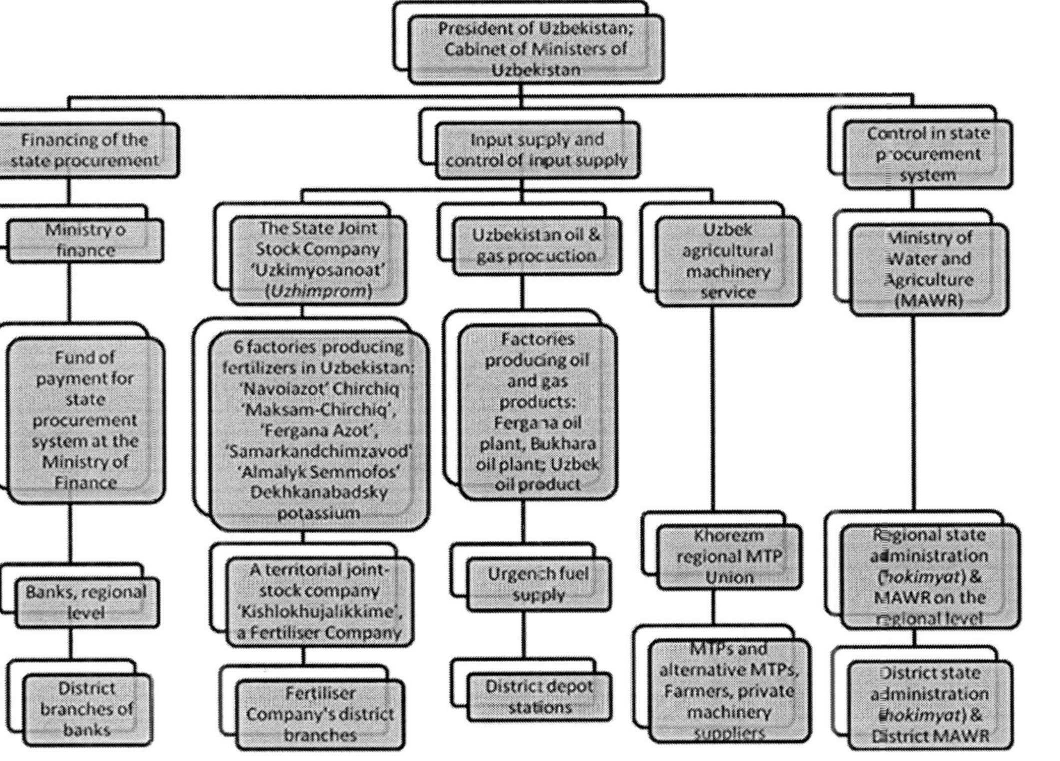

Figure 6: Management of state procurement system in Uzbekistan, 2009.

Source: Adapted from Niyazmetov, personal communication.

As illustrated in figure 6, there is control over agricultural production from the national to the local levels. The Cabinet of Ministers is responsible for the fulfilment of agricultural goals at the national level. At the regional and district levels, the MAWR and state administrations (*hokimyats*) exert control over agricultural production. In addition, the Fertiliser Company, MTPs, MTP regional Union, regional banks and fuel supply play a role in the fulfilment of state agricultural goals. Government control of agriculture is discussed further in Chapter 4.

At the farm level

At the other end of the state procurement system are the farmers. They face a number of difficulties in fulfilling state orders. One main problem is a lack of finance for production. According to the law, a subsidised loan is given to farmers at the rate of 60% of the cost of contractual agreements for growing cotton and wheat. Agrobank is the main bank for farmers (Information Portal about banks of Uzbekistan 2012). As indicated by the following quotes, the rest of the costs have to be covered by the farmer alone.

'It is up to farmers where they will get the inputs' (Farmers' survey 2009, communication with ZEF/UNESCO Urgench employees, 2010).

'There is a misunderstanding: the state gives a subsidy for 60%, but requires 100% prepayment from farmers. A farmer has a huge hole in his pocket!' (Interview with officials at district branch of the F. Co, June, 2010) Farmers can receive the three per cent loan for inputs for growing cotton and wheat only via buying inputs from AGSOs and not from the private market. In this way, AGSOs and farmers are interlinked in the frame of the state procurement system. During this process of allocation of subsidised loans to farmers, a number of problems occur from the side of different AGSOs. Many inconsistencies within the subsidised scheme occur in regard to AGSOs. For example, farmers often do not pay AGSOs on time as the allocated subsidised loan is hardly enough to cover production costs. Also, the (state owned) banks are slow in transferring loans to the farmers. Farmers may also prefer not to pay and have reason to believe they can get away with it (farmers' survey, interview, observations 2009 and 2010). This is further facilitated by the fact that many farmers are not used to book-keeping. AGSOs have to do everything possible to ensure that the harvest of cotton is good. They do it by frequently providing services to farmers without being paid for them and by servicing the state in supervising farmers' fields and the cotton growing process (see Chapter 4).

Farmers face the following shortcomings and problems as a result of a number of reforms. The first shortcoming of the current agricultural system is that growing state-ordered-crops is unprofitable, but farmers have nowhere to escape to and continue to stay in it. The next problem is that payment for cotton will be transferred sometime after the harvest. That means that the farmer still has to wait for the money, but taxes have to be paid immediately! As a farmer has not yet received the payment for cotton, a debt on the monthly taxes is accumulated (Farmers survey, Shomohulum district, 2009). If the farmers were private decision makers for their production, they would try to select and find better inputs (i.e. better types of seeds, fertilisers and other inputs). Under the current circumstances of the state procurement system farmers are not private decision makers of the cotton and wheat growing process nor motivated for doing it, as indicated by two quotations.

'Nothing depends on us (farmers). We are not heard…What will be done or decided for us, will be done by the state and this is the way it will be' (Farmers of 5 interviewed districts of Khorezm, 2009).

And

'I am tired of being a farmer - there is no rest, persistent problems, and i do not get enough of sleep' (A farmer, Khorem region, 2010).

Currently, farmers have low motivation to work in agriculture and are discouraged. They have to cover the production costs with regard to the state procurement system by themselves. In comparison, during Soviet times, kolkhoz workers were financially and morally encouraged to pick cotton and fulfil the state plan. In terms of money for a good harvest of cotton, people received a '13th month salary', premiums, good salaries sufficient for living, buying a car, having a vacation with the family once per year (for more details see Section 3.2).

Another obstacle for farmers is a lack of agricultural background. Many farmers now used to be medical doctors, teachers accountants and in professions unrelated to agriculture (Kazbekov and Qureshi 2011). As part of the Soviet path dependency, not all farmers have learnt how to work, how to make decisions independently without orders from above, how to be profitable, how to manage their own expenses. To address this problem, there is state care and control on the initial state of independence introduced by the Uzbek government (see Chapter 4).

For additional income, farmers grow commercial crops in addition to state ordered crops. The most widely spread commercial crop grown by

Khorezmians is rice. Wegerich (2006) explains that a farmer grows rice not just to cover the production costs of the cotton, but also to have some cash for living, as payment for cotton from the state comes to the account of the farmer and this money will be automatically used to cover production costs. Moreover, the farmer has no access to his bank account. Even though cotton could be profitable, the system does not really allow the farmer to have access to his/her profits (interviews with farmers, Wegerich 2006). Hence, with non-state-order crops the farmer has the direct benefit of having access to money outside the banking system.

3.5 Transformation of agricultural service organisations: from servicing 10 collective and state farms to 100 farmers

In this section, I summarize the situation of AGSOs in Uzbekistan during Soviet times and describe the subsequent changes that occurred in the organisations during the agricultural reforms (see section 3.2).

The history of many AGSOs dates back to the 1930s, when the Soviet agricultural system was initiated. After the Second World War, service infrastructure was promptly developed along with the kolkhozes and sovkhozes. AGSOs were in their glorious times, when all organisational needs were covered by the government (see Figure 5). Anything needed, like seeds, machinery or spare parts and fertilizers, was requested by the head of kolkhoz and sovkhoz and supplied by the Ministry of Agriculture, which was responsible for the agricultural production and supply. Regarding financial issues, every AGSO had a contract with, the only one at that time, Central Bank and all financial issues were transacted via it. AGSOs' accountants did not have to take care of finances and debts; if any debts occurred they were written off and dedicated to occasions like the 60th anniversary of the October revolution. Insurance services were offered by one state company dealing with issues in agriculture. In terms of management, every kolkhoz/sovkhoz had a head (a director), an accountant, a veterinary sergeon, a zoo-technician and other specialists, which made it an independent production unit. Services like commodity exchange and consulting did not exist. There was no need for commodity exchange, since all products were ordered, produced and purchased by the central government. Consultants were not necessary either because there was a trained agronomist in each kolkhoz and sovkhoz who was competent in all production issues. According to Kandiyoti (2007), there were Consumers' Cooperative Associations that played a central role in marketing kolkhoz produce. Cooperative societies ran shops, catering establishments and rural industrial enterprises that produced canned goods, fruit juices and bakery items.

The end of the Soviet Union in 1991 caused political and economical changes, which had an impact upon AGSOs functioning and evolution. Independent Uzbekistan was cut off from the budgetary grant that it used to receive from the USSR (Khan 2005). The dissolution of collective farms spurred a host of successor organisations, at district and lower levels. The transition has been complicated by the heterogeneous lineage of AGSOs: some derive from kolkhoz sub-organisation, while others are descendents of district or regional organisation (as exemplified by my cases). Some services have emerged due to the present needs of the farmers that had not been covered before (Khan 2005; Van Assche et al. forthcoming). State support of service providers reduced dramatically between 1991 and 1994. The lack of state support as well as an end to receiving orders, plans and assistance from the state in terms of production and management, was a crucial milestone in the evolution of AGSOs. The availability of established infrastructure for agricultural service provision, which was meant for kolhozes/sovhozes in Soviet times, was the basis to develop service provision with regard to the needs of the current agriculture and its clients (farmers with land size up to 200 ha and rural households with land size up to 1 ha). One more negative impact of the breakdown of the Soviet Union on the agricultural sector was a breakdown of links between former Soviet republics. Production of agricultural inputs and machinery parts took place in Russia and other post-Soviet republics.

A turning point in the evolution of service providers during transition came with the process of individualization of farmers in 2004. AGSOs moved from being centrally managed and providing services for a few state farms to providing services to a much larger contingent of individual farmers (Shtaltovna et al. 2011). Service providers have transformed quickly since 2004. AGSOs were made financially autonomous overnight, without preparing them or giving them the necessary assets and infrastructure, while still holding them to a series of obligations stemming from the state procurement system for cotton and wheat (Van Assche et al. forthcoming). Before Independence, AGSOs were subordinate to the government and instructed on their role in fulfilling state plans. Today, they remain subordinate to the government but have to act independently and be profitable (Interview with the MTP economist, 2009). 'Now we subordinate to all and to no one at the same time' (Interview with a manager of repair workshop, MTP, 2009). Although large farms have been reorganized into private farms within a short period of time, re-organisation of services for newly emerged farmers will require a long time and substantial funds (Jooshev and Mityakova 2008, interviews).

Further land consolidation processes in 2008 made the management and control over cotton and wheat producers easier for the state. For AGSOs, the inherited debts of the farmers whose land was consolidated, poor condition of the service and input provision, inability to renew hardware, superfluous meddling in the business of service organisation and farmers, and pressure from the state to fulfil the state plan, brought a lot of stress and disappointment (Van Assche et al. forthcoming). As a result, many AGSOs are at the brink of bankruptcy in recent years. 2009 was the next step in agricultural reform chains. Lands were consolidated again but to a lesser degree than in 2008. Farmers who performed better were given the lands of those who did not. With that reform, the central government introduced 100% prepayment by farmers to service providers under the state procurement system. It was an attempt to respond to numerous problems that occurred between farmers and AGSOs[12].

The erosion of rural livelihoods in Uzbekistan, as in the rest of Central Asia, must be understood as a result of the decay of an ensemble of institutions involved in production, distribution, and service delivery (Kandiyoti 2007). Changes related to supply and managing of AGSOs that were started in the early 1990s are still not properly addressed and not yet solved. Nevertheless, the fine-tuning of the system is taking place every year and the service infrastructure has been adjusting up to now to serve farmers.

3.6 Overview of agricultural service providers in Khorezm

In the light of the historical background sketched above and the changes introduced in agriculture in the independence period of Uzbekistan, this section will focus on the recent status of service provision in Urgench. Not much has been written on AGSOs in Uzbekistan or Central Asia except the agricultural services' overview by the NBT (2008), a master thesis written in the frame of the Project by Niyazmetov (2008), Sehring (2005) and Kandiyoti (2007). Therefore,

12 A further detailed account of the transition of different AGSO over time can be found in Annex 4 'Structural changes in agricultural service providers: From Soviet period up to now'.

this section is mainly based on data collected through the farmers' survey I conducted in 2009[13].

The major services available to agricultural producers in Khorezm province of Uzbekistan are presented in Table 1. They are provided at the district level by water user organisations (WUAs), alternative machine tractor parks and machine tractor parks ((A)MTPs), fuel suppliers, the Fertiliser Company, banks, veterinary stations, and bio-labs. Other emerging services for farmers are not presented in Table 1 but will be discussed later in this section.

Water supply in Khorezm is organised through Water User Associations (WUAs). There are a total of 112 WUAs, which were established on the basis of previously managed collective farms (Zavgorodnyaya 2006; Veldwisch 2007; Oberkircher et al. 2010). WUAs are defined as associations (unions) of the newly formed farmers' and other legal and physical persons carrying out economic and other activities related to the selection, use and discharge of certain amounts of water. WUAs are established to ensure the regulation of water relations in the transformed agricultural enterprises (Cabinet of Ministers of Uzbekistan, № 8 of 05.01.2002). Due to the insufficient funds allocated to maintain the irrigation system, the state decided to establish WUAs as a main managing body for delivery, distribution and allocation of irrigation services as well as maintenance of the irrigation system. WUAs function according to the Law on 'Water resources and water use' adapted in 06.05.1993 # 837-XII. Since they were established in 2001, there have only been minor reforms introduced in the regulation of WUAs. One exception is the change in 25.12.2009 #240 that concerned the reforms in agriculture and title of Water User Association changed into Water Consumer Association (WCA). Apart from legal reforms, there have been attempts to support WUAs from local, district and regional government by providing office space, computers as well as providing employees in order to facilitate their activities.

---

13 This chapter builds mainly on the results of the conducted farmers'survey in five districts of Khorezm province, Uzbekistan. For more details see methodology section in chapter 1. Also, see annex 4 on Structural changes in agricultural service providers: From the Soviet period up to now'.

Small farmers face problems and conflicts related to water distribution mechanisms (farmers' survey, 2009). Rural households (*dekhans*), being a rather poor category of the rural population, face a lot of problems in getting water because electricity is often switched off and irregularly working pumps. If farmers have

Table 1: Number of district branches of the agricultural service providers in the Khorezm region

District	AMTP	WUA	Fuel suppliers	Fertilizers	Banks	Vet services	Bio-labs
Bagat	12	11	9	6	14	15	12
Gurlen	12	11	12	7	10	16	13
Kushkupir	15	17	14	14	11	19	13
Urgench	13	13	14	12	18	17	9
Khasarasp	13	12	10	8	11	21	11
Khonki	11	11	12	10	12	14	10
Khiva	11	10	10	8	11	16	8
Shavat	14	12	10	8	15	19	14
Yangiyarik	9	9	9	9	8	13	8
Yangibazar	6	6	9	10	9	14	10
Total	**116**	**112**	**109**	**92**	**119**	**164**	**109**

Source: Regional representative of Ministry of Agriculture and Water Resources of Uzbekistan (2010), based on data from 01.07.2009.

problems and questions related to managing their agriculture, the directors of the WUA are knowledgeable people that they can go to for advice. The main problems WUAs face are the lack of qualified experts, insufficient financial resources, poor technical base/equipment, and electricity supply. In-depth studies on water allocation, distribution, process ownership can be found in other publications (Zavgorodnyaya 2006; Veldwisch 2007; Yalcin and Mollinga 2007; Abdullaev et al. 2008; Veldwisch 2008; Abdullaev and Mollinga 2010; Hornidge et al. 2011; Oberkircher and Hornidge 2011).

In contrast to water provision, multiple sources of **machinery** can be found in the Khorezm region. Primary sources of machinery are Machine-Tractor-Parks, which previously belonged to the collectivized farms and were reorganized into joint stock companies in 1997 (Uzbekistan). Additionally, 'alternative' MTPs (AMTPs) were founded in 2004 on the basis of the earlier tractor parks that existed in every *kolkhoz* and *sovkhoz*. MTPs and AMTPs are regulated in accordance with the state law 'On Joint-stock Companies'. Issues related to machinery providers (MTPs and AMTPs) that concerned farmers were ever increasing prices, long wait time for machinery and outdated equipment. There is a lack of harvesters and long queues during the season. Machinery is given first to the farmers growing state ordered-crops and to those who have links to the state. The poor quality and unreliability of the machinery from the MTPs were also reported by Trevisani (2008).

In addition to the MTPs, the Fertiliser Company, the cotton factory, and many farmers possess their own machinery. They use it for their own purposes and sublet to other farmers and *dekhans*. Farmers who use private machinery indicated that the service was better than from the MTPs. In the survey, they reported that the privately owned machinery is newer; their providers are more flexible in cooperation and charge lower prices; problems with the machinery could be reported to the provider; there is less paper work; and the machinery is provided faster than by state MTPs. Apart from the MTPs and AMTPs, *dekhans* can find machinery at the fertilizer company, the private service providers, the neighbours, friends, and the farmer they are working for. Thus, there are several options available to obtain machinery (see Section 3.7.2 for more details).

The fuel supplier maintains a monopoly status in Khorezm region (and in Uzbekistan) and it functions based on the Presidential order #UP 470 from 07.09.1992 on 'Measures of bringing the order in sales of petroleum products and improvement of management of petroleum branch of economy'; Resolution of Cabinet of Ministers # 523 from 15.12.1998 on 'Issues of organisation and functioning of national holding company 'Uzneftegaz'; a legal document #289 from 28.06.2003 on 'Approval of licensing of sales of petroleum products'; Resolution of Cabinet of Ministers # 523 from 15.12.1998 on Issues of organisation and functioning of national holding company 'Uzneftegaz'. A regional network of filling stations has been developed since 2004 and thus became more accessible for farmers than before when farmers had to travel to the district centres. As with other service providers, farmers growing state-ordered-crops enter into a contract, pay the money (from the subsidized loan) and after that, according to the allocation schedule, they will get the fuel. Additionally, fuel is available through commodity exchange and some private filling stations. The indicated issues of the fuel supply are that farmers pay too much and they do not receive the value they pay for. Apart from this, farmers are satisfied with service provision by the fuel spply company. Other types of farmers and rural households have access to the same sources of fuel supply, the only difference is that they do not have a subsidized loan from the state.

Financial services are provided by banks and work according to the Law # 216-I dated from 25.04.1996 on 'Banks and banking functioning'. 'In the Soviet era, banks were largely accounting centres, an epiphenomenon of the planning system, but now they became a fulcrum of transition' (Burawoy 2009). For the farmers growing wheat and cotton the subsidized loan (three per cent annual rate), allocated mainly by the "Pakhta-Bank" (for the cotton-growing sector) and the "Galla-Bank" (for the grain production sector), which are fully controlled by the government (Ilkhamov 2000). Farmers dealing with poultry, animal husbandry, horticulture, and silkworm breeding do not have any subsidized state loans. The credits for farmers are available at the regular rate, i.e. 16-18% thorugh the state owned banks. Farmers as a production unit were established in 2004-2005, and the banks do not have a life-long experience in serving individual clients. In addition, the relatively complex scheme of credit allocation, which is changing every year, generated a number of serious complains from the clients towards the financial service provider. The most often mentioned difficulties by farmers in their encounters with banks are that banks work slowly, transfer of money to the service providers is delayed, which causes delay of service provision (which is crucial in agriculture), banking requires a huge amount of paper work, there is no money available in cash or to withdraw, and all income has to be put in the bank. Moreover, while taking an expensive commercial loan, a farmer has to give collateral even though he has nothing of value. *'Land does not belong to me, my house cannot be put as a mortgage, what else can I deposit?!'* (Farmers' survey, 2009). All farmers indicated the necessity to bribe the bank workers in order to speed up the procedure of money transfer or getting a loan. All cash transactions go through the bank, so the state can keep track of farmers' cash and money flow, and prevent inflation and corruption. This is how the state justifies control. There are no institutions like credit unions that would offer alternative sources of money to farmers and rural households. Farmers who neither get a subsidized loan nor have enough money for commercial credit borrow money from friends. There are not many options for rural households to borrow money from the bank or they are not aware of possibilities.

There are six factories in Uzbekistan producing all required **fertilisers** for agricultural production and thus Uzbekistan can fully supply farmers with needed inputs. These factories are: Navoi factory 'Navoiazot' (nitrogen fertilizers, nitrate), Chirchiq 'Maksam-Chirchiq' (nitrogen fertilisers, urea, sulfate), Fergana factory 'Fergana Azot' (nitrogen fertilisers, phosphoric fertilisers), Samarkand factory 'Samarkandchimzavod' (phosphoric fertilisers), Almalyk Maksym factory 'Almalyk Semmofos' (phosphoric fertilisers) and Dekhkanabadsky potassium plant. These six factories producing fertilisers together with the Fertiliser Company (F. Co) responsible for supplying fertilisers to farmers growing cotton and wheat are subordinate to the State Joint Stock Company 'Uzkimyosanoat' (*Uzhimprom*). The F. Co is represented in all regions of Uzbekistan with its regional office and regional branches. From 2004 to 2010, the amount of prepayment for fertilisers for farmers growing state-ordered-crops has increased from 30 to 100%. Farmers are satisfied with the supply of fertilisers, which are delivered directly to the fields escorted by the local police, prosecution officer and the representative of the local branch of the fertiliser organisation. The regional branch is well developed, thus farmers do not have to travel far for the input. However some drawbacks in regard to fertilisers supply were indicated, such as high price (yearly increase), the allocated amount of fertilisers under the subsidized scheme is unsufficient, farmers have to purchase the missing amount through commodity exchange or private shops.

Apart from the F. Co that is the main supplier of farmers growing state-ordered-crops, there is a growing number of alternative options to purchase fertilisers (Khan 2005). For instance, one can buy fertilisers from the F. Co through entering into a contract, commodity (produce) exchange (birja), district stocks of fertilisers (fertilisers from the factories), regional network of shops[14] subordinated to the *Uzhimprom*, and bazaars. Farmers who get sufficient income and those operating outside the state order system prefer to buy fertilisers from alternative sources owing the fact that they supply a 100% pure product, are more flexible in cooperation, require less paper work, and the product is always

14 The Ministry of Chemical Production, which is the apex standing organization of the Fertilizer organization, has a subsidiary unit in the Khorezm region. It deals with commercial sale of fertilizers. It has 46 shops (32 are functional) in the region, selling fertilizers (Interview with the official at the Fertiliser Company, 2010).

available. As regards rural households, they buy fertilisers from private shops and the regional network of shops and markets, and use natural manure, which can be bought from neighbours and friends (interviews in *Uzhimprom*, Tashkent, 2010). Thus, the private market of fertiliser is slowly evolving. In contrast, the Fertiliser Company works mostly to provide fertilisers for farmers growing cotton and wheat. Nevertheless, the F. Co remains the main supplier for farmers under the state procurement system (see Section 3.7.1 for more details).

Bio-laboratories (bio-labs) came into existence in the early 1980s, partly to promote and organize organic pest-control and partly as a way to save on pesticides (Van Assche et al. forthcoming, interview with the bio-lab director 2010). Some bio-labs are private companies while others are joint-stock companies, with a majority of state shares. The latter types were organised based on the previous Soviet hierarchical system and subordinate to the regional and national plant protection organisation, which is the mother organisation, whereas the private ones have emerged in recent years. Usually, there is one bio-lab per village. Some farmers complained about the quality of products of the state bio-labs. Having quite a choice of bio-labs, farmers can choose with which bio-lab to enter into a contract based on quality of bio-methods and service delivery. The problem occurs when all farmers need bio-methods at the same time and then bio-labs cannot supply all simultaneously. In addition to the bio-methods, there are foreign pesticides and herbicides advertised on the local TV, which can be obtained in particular private shops located in district centres, and thus far for some farmers to reach. While dealing with the representatives of foreign organisations, farmers cannot use any informal connections to get the input for free or without queuing because this is a private organisation. The survey showed that it is usually expensive for *dekhans* to purchase bio-methods. Therefore they try to get in groups and buy it together or by appealing to informal arrangements like offering paper, sugar, and flour, which are inputs needed in the bio-lab. Usually, the bio-labs' directors are kind to dekhans and provide them the product at the suggested condition or for free (see Section 3.7.3 for more details).

In addition, bio-lab services were supported by the GIZ GmbH regional programme 'Enhancing Economic and Environmental Welfare in the Aral Sea Region' (EEWA). Value chain promotion is a key component of the program (Serving the People of Central Asia). The value chains are set so as to generate explicit environmental benefits, for example, by applying principles of sustainable resource use and management in agriculture, by introducing integrated pest management (IPM) - including the use of so-called beneficial organisms, and by adopting product quality standards. The latter ensures enhanced consumer and environmental safety. It also helps to secure established markets and to open both new regional as well as global markets for higher value-added products. The value chain promotion focuses on developing the potentials of the following sub-sectors: sugar melon, greenhouse products, fodder crops, livestock, aquaculture and fisheries (Serving the People of Central Asia).

Veterinary service (vet-service): Veterinary service in Uzbekistan functions based on the Presidential decree from 23.03.06 №308 'On measures to stimulate the quality of cattle in the private and state farms. This decree also foresaw establishment of veterinary points in every village. Depending on demand and size of the population, two or three vet points could be established (Interview with the Head of the Regional State Veterinary service, Urgench, 2009). Thus, there is a veterinarian in every village, available from the state and private providers. A farmer can appeal to the one (s)he prefers. Some farmers refer to private ones, saying that their rates are lower, they can be reached 24 hours/day, they are more flexible in work, and the vet's experience is larger; a private vet can visit the farmer's house and check how things are going. Some farmers have never heard of the existence of the private vets. Many are satisfied with the public vets. S/he comes only twice a year to give an injection to the cows (farmers' survey 2009). There are cases when a veterinarian works as a private and a public vet.

Emerging services

Insurance is a relatively new service in agriculture. There are state and private insurance companies. However, insurance is not commonly used in agriculture. Insurance is not an obligatory service for farmers growing state-ordered-crops. Being not a common service in Soviet times, insurance nowadays remains mysterious to farmers. The main farmers' feedback on insurance is that many farmers do not really get the idea of insurance, they are not sure that insurance can pay the insurance cases; they do not know how to claim for paying the insurance cases, or heard from others that money cannot be paid back in the insurance cases (Farmers' survey 2009). Insurance organisations do not conduct information or advertisement campaigns to attract clients and explain farmers the benefits of insuring their crops and business. Therefore, this service (or business) remains to be fully developed in Khorezm.

Commodity (produce) exchange (Birja): The first commodity exchange appeared after the dissolution of the Soviet Union, in 2004-2005, with the idea to provide agricultural inputs for commercial farming of the farmers and dekhans. Commodity exchanges function on the basis of Law on 'Commodity exchange and commodity exchange activities' #625-XII issued in 1992. As time has shown, commodity exchange also provides the missing inputs for the farmers growing state ordered-crops. A commodity exchange sells products at a similar price to prices by the AGSOs at the 100% prepayment condition. Fertilisers, fuel, forage for animals can be purchased via the Commodity exchange. There is a network of shops through the region by the Commodity exchange where the rural population can buy a bit more expensive but still needed inputs for farming. This is a relatively new service provider and it will take time before it gets established.

Information and consulting service: Applied agricultural research (part of which could be labelled as extension services) was in Soviet times vast, complex, and evolving (Morgounov and Zuidema 2001). Applied research institutes and experimental farms existed everywhere in the Union. There is an association of farmers which is de jure supposed to provide information and consulting services to farmers (Kandiyoti 2007) . However, the results of the farmers' survey showed that farmers hardly ever use and have not heard of the services of this organisation. Moreover, it showed that farmers very often lack knowledge on how to enter into a contract, the matter of insurance, how to defend their rights while working with service providers, when new pests appear in the gardens and there is no one to ask for agricultural knowledge how to deal with them. To fill the gap in service and specialists, farmers (especially those growing state-ordered-crops) usually appeal to the representatives of the state administration, i.e. the head of village/town council, other representatives of the state administration and the district Ministry of Water and Agriculture, the Land surveyor, TV programmes, and the journal 'Uzbek Agriculture'. Some, especially hokims, can directly contact the heads of AGSOs (MTP, bank, and Water Consumer Associations) depending on the issue concerned), and farmers will get the required answers and assistance. The heads of bio-labs, MTPs, and vets were specialists in their respective fields since Soviet times. Therefore they are usually approached by farmers directly when they have some problems (Wall 2006; Hornidge et al. 2011). Thus, at present directors and managers of AGSO and state administration who used to be kolkhoz employees are knowledge carriers, and their expertise in agriculture is especially appreciated by farmers when there are no alternative sources of knowledge and information. In addition, there are specialists on different issues in local administration (animal husbandry, etc.) who can give advice. Also, to get answers to questions or assistance in regard to some services, farmers often use their personal networks and kinship relations (Farmers' survey, 2009, Oberkircher 2010). Farmers engaged in horticulture, animal husbandry, and poultry production and getting less support from the state representatives, showed a more active attitude and independence while searching for answers to their questions. They request private service providers and appeal to the court in cases of disagreement, in contrast to state crops' growers.

Services that farmers and *dekhans* indicated as lacking: supply of good quality of fodder for animals, more varieties of vegetable and plants seeds, fruit and vegetable processing plants (which used to be there during the Soviet times, but no longer exists) and more diverse sources of products and inputs. Seeds for cotton and wheat are available from the cotton and wheat gins but were referred as rather expensive and the quality remains the same, that is does not improve over the years. Non-state-crop-growers, being outside state attention, became primary users of the private service providers such as bio-labs, outlets and private shops for fertilisers, using commercial credits, and the private vet (Evers and Wall 2006; Turaeva-Höhne and Hornidge 2012). In addition, trying to fill the gap of the missing services and find better quality products they are getting inputs from the markets and from other countries. For instance, some farmers go to Russia to buy chicken, or purchase spare parts from Turkmenistan. *Dekhans* remain with the least attention from the state. They cannot appeal to the *hokimyat* and there is no alternative information source for them. Nevertheless, they try to find a way out. For instance, they talk to old specialists, other farmers, and read books. There are new services emerging, services which were not there before or outside the attention of the state due to the fact that they have little to do with the state procurement system. Branches of international organisations like Ifoda (herbicides and pesticides), and international machinery suppliers are examples of organisations function based on market rules, i.e. are money-based, require 100% prepayment, and have nothing to do with state control, thus there is different attitude towards running a business.

Summarizing this section, we can see that change and fine-tuning of the rural economy are still taking place. Such service providers as MTP, bio-laboratories, water supply, veterinary services are there since Soviet times are undergoing changes. The AGSOs described in the first part of this section mainly evolve based on the previous, Soviet system of service organisation. Moreover development of most of the agricultural service providers is strongly related to the dynamics in the state procurement system. Due to the poor functioning of many of AGSOs, they do not have financial resources to renew their existing infrastructure.

Some services were indicated by farmers and rural households as missing due to the fact that the state administration does not take care of them as it used to be during the Soviet times. These are good quality of fodder for animals, more varieties of vegetable and plants seeds, fruit and vegetable processing plants. To fill this gap private service providers slowly emerge and try to meet the farmers' needs. Private service provision is most interesting for farmers who grow crops outside the state procurement system or engaged in animal husbandry. Moreover, those farmers showed to be proactive in looking for missing services. Newly emerged services usually work according to market principles, in contrast to agricultural service providers that have been established since Soviet times.

3.7 Introducing the three case studies

In this section, I take a closer look at three agricultural service organisations in Khorezm: the Fertiliser Company (at the regional and local levels), a private bio-lab, a machine tractor park and a newly established private pesticides' provider Ifoda. As indicated above, I analyse the evolution of these organisations in the light of agrarian change[15].

3.7.1 A Fertiliser Company

The Fertiliser Company (F. Co) in Uzbekistan is more centralized than most other AGSOs (like the MTPs and bio-labs). The F. Co, together with the fertiliser factories are subordinate to the State Joint Stock Company 'Uzkimyosanoat' (*Uzhimprom*). Under the name 'Kishlokhujalikkime', the Khorezm F.Co is a territorial joint-stock company that functions at the regional level. The stock-holders of the F. Co are *Uzhimprom* and regional and local subsidiaries. At district and local levels, the organisation functions via a network of 11 district branches and 92 distribution points[16] (see Table 1).

The F. Co is supposed to be a self-financing organisation, covering the costs of its regional office and the 11 branches by selling fertilisers and paying out yearly dividends (6%) to the shareholders. At the regional office, 38 people are employed, and 776 people work in the district branches and distribution points.

15 I would like to acknowledge that the case descriptions were prepared and used for a joint publication with Kristof van Assche and Anna-Katharina Hornidge and the author (van Assche et al. forthcoming).

16 The internship for this case study took place mostly at the level of the regional management organisation, but I also had the opportunity to visit local distribution points and district branches and talk to people there. Farmers and people working at other AGSOs were interviewed regarding the activities of the fertilizer company. In the following, I will refer to it as the Khorezm Fertilizer Company, or F. Co.

The main goal is the management and control of the fertiliser supply for the state ordered crops, cotton and wheat. These branches and points replaced *kolkhoz* fertiliser storage, often on the sites of the old storage facilities.

3.7.2 A Machine Tractor Park

From Soviet times up to now, machine tractor parks (MTPs) remain vital for the growing of agricultural crops, as is its human capital for present day management in the command agricultural system. The history of MTPs in Uzbekistan is quite convoluted. The history of the Urgench district Machine-Tractor-Park tracks back to the 1930s, when the Soviet agricultural system was initiated. At that time, the MTP was established on the district level and supplied kolkhozes with machinery. After the second World War, in the 1950s, when machinery became more common, the kolkhozes got their own parks and the earlier MTPs served as repair stations, where machinery from all kolkhozes was brought and repaired. The 1960s and early 1970s were the boom period for the Urgench MTP, when it was responsible to supply all kolkhozes of the Urgench district with equipment, machinery, spare parts, coal, fertilisers, and seeds. Kolkhozes did not have to purchase any input, they were fully supplied by the MTP. In addition, a land reclamation department was established at the MTP for developing new lands in the district. Further on, in the 1980s, when the

Figure 7: The MTPs were established in 1930s with the initiation of kolkhozes and sovkhozes. Photo: A. Shtaltovna 2009

lands had been developed, the land reclamation department at the MTP was closed down as well as agricultural supply function was moved to other organisations and so the MTP continued to deal just with supply of machinery (ploughs, sowing machines, and spare parts) for agricultural purposes. During the period of transition (beginning to mid-nineties) MTP dealt with repair of agricultural machinery and production of spare parts. In 1997 (Uzbekistan) MTPs were reorganised into joint stock companies. Additionally, 'alternative' MTPs (AMTPs) were founded in the late 1990s on the basis of the earlier tractor parks that existed in every kolkhoz and sovkhoz. MTPs and AMTPs are regulated in accordance with the state law 'On Joint-stock Companies'. As machine-tractor parks are the main machinery service providers to farmers in Uzbekistan, the aim of these reforms when they were first introduced was clearly to strengthen the material and technical basis by increasing their role in productive and technical service provision to agriculture. The supply of machinery and spare parts, previously received from the Regional MTP Union, stopped between 1991 and 1994. The few spare parts that were still received had to be paid for and thus were no longer supplied free of charge. In that period the spare parts started to randomly appear on the open air markets in Urgench and were brought by individual merchants from neighbouring Turkmenistan. The resulting poor financial situation of the MTPs has led to a steady decrease in their numbers in the Khorezm province in the past years, counting 13 MTPs and 196 ***AMTPs on July 1, 2010 (Regional Statistics Office 2010).***

Currently, the Urgench district MTP is a joint-stock company with the government's share being less than 35%. The main functions of the MTP are (a) rendering mechanical services, (b) repairing agricultural machinery, and (c) the production and supply of spare parts. The MTP's clients are the farmers of the Urgench district, state organisations, agricultural enterprises, husbandries, district AMTPs and other organisations.

In addition to serious financial problems of the MTP, mainly caused by the state agricultural policy, there is rapidly growing competition in providing machinery services in the Urgench district. Apart from MTPs and AMTPs, machinery is available from a number of farmers and a fertiliser company. Competition is growing mainly because farmers now have more money than in the past and clients are not always satisfied with the machinery from the MTP in regard to price and quality, particularly as they often receive worn out machinery.

3.7.3 A Private Bio-laboratory

The private bio-lab I focused upon is Aek Durman. It continues to work in a Soviet style, either by thinking in terms of planned service delivery, or by think ing in terms of its contribution to the agricultural production system, rather than individual profits. Before 2003, the year the bio-lab was established, its director used to work in one of the first labs in the region, established in 1983. It is in practice a one-man organisation. The director is very enthusiastic, in love with his work, and he wants to 'save cotton from the pests'. A slogan of this bio-lab director is 'the best farmers will find the best bio-labs, and vice versa. They will find each other'. The director describes himself as a communist. According to himself, he is not interested to make money, he protects inter ests of farmers, sells them cheap, asks other labs to work at affordable rates. He likes people who work hard and are inquisitive, since better crop protection methods can be discovered by good thinking.

The bio-labs have a large degree of autonomy, as they fill a specialized niche, cannot be replaced easily, and do not represent strong financial return to the central government as in the case of Fertiliser Company. In other words, they can do what they want, since nobody is interested in them. This autonomy allows for business-like thinking, and profit optimization. Bio-labs could, in theory, evolve faster than other AGSO under stronger state supervision, be cause they do not require a lot of investments, land, infrastructure, personnel, and could make use of a rapidly changing science. Additionally, they are not as extensively involved in the state agricultural campaign as MTPs and the F. Co are, which again allows them to spend more time on their business venture.

Yet, access to new science is difficult, and, if farmers do not pay for the services, it is hard to think of even minor investments in innovation (Hann 2003). A wide variety of innovation and adaptation paths were observed, related to differences in management, ownership and in the local community.Once bio-labs have a piece of land, many of the necessary inputs can be produced on site. Bio-lab equipment is usually there, a Soviet legacy. New lab equipment is rare and very expensive.

3.7.4 Ifoda: a pesticide and herbicide provider

As part of the field research, I looked at a newly established service provider. Ifoda has no roots in the Soviet past and is legally a private entity. It is an international company that sells pesticides and herbicides. Ifoda has a good financial position. It is not linked to the state procurement system and it receives international expertise how on working procedures. Ifoda sells products at only one condition – cash. The structure of this organisation and its working style are based on market principles.

The relatively short case descriptions set the stage for the case analysis, organized around a series of concepts found to be helpful in explaining the evolution of these organisations. In the following empirical chapters, I will discuss the functioning of each organisation through the lens of its interactions with their internal and external environments.

Figure 8: A bio-laboratory.
Photo: A. Shtaltovna 2010

3.8 Conclusion

This chapter introduced the reader to a chain of agricultural reforms in Uzbekistan since Soviet times up to today and the evolution of AGSOs through this period. Based mainly on my empirical data I presented a close look at the processes of agrarian change and up to date status of AGSOs in Khorezm. In particular, I looked at the evolution of AGSOs in times of agrarian change, the implications of previous reforms on the present day agriculture in regard to AGSOs and farmers. This data is a basis for the empirical chapters.

Based on the chain of agrarian reforms and AGSO in it, I draw on the agrarian structure's path dependency in the present day agriculture in Uzbekistan and the broader theoretical framework of the Technical and Administrative Task Environment (TATE) (see Section 2.2).

There is a path dependency in agrarian structure in the present day Uzbekistan. The central state applies old Soviet methods of dealing with issues of present day agriculture. Amongst others, these are: giving plans for cotton growing, allocating subsidies, involving state administration in supervising cotton growing and involving all population in cotton picking.

Building on this, in the following empirical chapters I continue the discussion on agrarian change and transition, where we will see the implications of discussed reforms in agriculture on the daily basis of AGSOs functioning in present day Uzbekistan.

CHAPTER 4: ALIGNMENT WHILE KEEPING DISTANCE: THE NEGOTIATION OF RELATIONS BETWEEN AGRICULTURAL SERVICE ORGANISATIONS AND THE STATE

4.1 Introduction

State–producer relationships are at the heart of agricultural production in Uzbekistan. According to Trevisani, the Uzbek model of state-society relations is one in which 'authority and state power are territorially defined, hierarchically conceived, and centrally organised' (2008). Furthermore, Uzbekistan's state-producer relations are characterised by 'blurred boundaries between state and society, by a weak participation of societal institutions in the public sphere, and by a submissive attitude of a social domain towards the domain of the state' (Trevisani 2008). Kandiyoti has argued that a political economic framework on the problems and survival strategies of primary producers can provide better insight into state-society dynamics in Central Asia's states than dependency or postcolonial theory approaches (2002 in Trevisani 2010). 'Market reforms have brought about a partial modification and restructuring of these institutions, retaining many features of the command economy, but 'hollowing out' the roles and functions of the organisation that were vehicles of social welfare and entitlements' (Kandiyoti 2007). In this chapter, I discuss state strategic interest with regard to cotton growing and how it mobilises other actors for that purpose. I shed light on the political economy of agricultural production[17] in Uzbekistan by examining the interactions between the state and three service providers in this chapter (see Section 2.7 for more details)[18]. My findings provide insight in the role of these different AGSOs in the state procurement system, and thus their importance to the central government during the transition period.

In Uzbekistan, the present day agrarian structure builds on Soviet collective and command system. Previous studies on the relationship between agricultural

---

17 For a more detailed analysis of theoretical perspective adopted in this chapter, see in Chapter 2.

18 In this chapter I refer not just to one studied bio-lab, but to 'bio-labs'. With that I mean that the evidence provided is based on the answers provided by the directors and staff of more bio-labs.

service organisations and state organisations indicated that AGSOs are not independent producers in the current system, despite the fact that the system is proclaimed to be 'moving towards market economy' (Karimov 1995; Wegerich 2006; Trevisani 2008). In agreement with these previous studies, I found that two of the AGSOs I investigated have continued to function as instruments of the government and their activities are controlled to a large extent. However, I also provide evidence that some are being left to develop with less attention from the state. I demonstrate that the relationships of the three AGSOs with the state differ in the following ways: frequency of interaction with the state administration; responsiveness to the state's orders; and the use of Soviet rules at work. I show that the degree of independence of each AGSO depends on the organisation's relevance to the state procurement system. Subsequently, path dependency (see section 2.4) is most pronounced in AGSOs most relevant to the state procurement system.

This introduction is followed by section 4.2 where I shed light on control measures employed in Uzbek agriculture and its integration in the present day procurement system in agriculture. Sub-sections 4.2.1, 4.2.2 and 4.2.3 reveal how the Fertiliser Company, the MTP and the Bio-lab are mobilised by the state for the state procurement system. The chapter closes with a conclusion in section 4.3, where I present the first components of a relational typology of AGSOs based on data from this chapter.

4.2 Agricultural service providers' relationships with state

'Economise the usage of white paper while working'

(The order from the Cabinet of Ministers of Uzbekistan to

the state relevant organizations, 2010)

During Soviet times there was an expression 'Cotton is the white gold of Uzbekistan, grown by the golden hands of the people'. This phrase has not lost its meaning up to now. Cotton remains the most significant agricultural crop[19] (see Annex 5 for agricultural production indicators in Khorezm region). However, after independence, budget allocations for cotton and wheat have changed. The central government wants to extract maximum benefits from the cotton and wheat sector, but it cannot allocate as much money as the central government of the Soviet Union used to. Therefore, the central government employs a vari-

19 After independence wheat became also an important crop in agriculture too, see Chapter 3.

ety of methods towards agricultural producers and service providers to prioritize cotton and wheat growing. So far, the agricultural system has maintained the hierarchical organisation introduced during the years of the early Soviet Union (Trevisani 2008). The government uses resources at its disposal, i.e. former *kolkhoz* leaders and the rural population, who still have the knowledge of how to work under the command system (Wegerich 2005; Wall 2006; Trevisani 2008; Wall 2008). However, there is less financial encouragement as compared to Soviet times (Ilkhamov 2000).

According to the official in the Fertiliser Company, the state protects and supports AGSOs and farmers, teaches them how to work in the new conditions in a protected environment (2010).

> 'Peace, order, life of people is important that's why there is a control on the initial state of independence' (An official in the F. Co, former governmental official, June 2010).

In this section, I demonstrate that there seem to be other reasons for controlin agriculture besides 'guarding farmers from the unknown to them environment' (interviews with the state representatives, 2010). The nature of economic transition in Uzbekistan can be described essentially as a deliberate attempt to concentrate all economic control and activity in the central government, which directly supervises a gradual and carefully calculated transition to market economy (Luong 2002). Due to the importance of agriculture, state control is a part of the current agricultural system in Uzbekistan and thus it is a part of agrarian structure's path dependency. While in Soviet times the communist ideology supported the state's call to join in the cotton harvest, this missing embedded ideology today is compensated for by the exercising of state control[20]. Mobilisation does not necessarily involve participation in decision-making; it may be driven by economic incentives or coercive political pressure (Fox 2007). The latter option is the one used by the Uzbek government, where the Uzbek government appeals to variety of methods such as meetings, visits to the organisations, conducting supervisions of the farmers' fields, *selector* meetings and others, which I am going to discuss in the following.

Laws pertaining to agricultural service providers

In order to control agricultural organisations related to cotton and wheat production, there are a number of legal documents issued by the central and local

---

20 This work does not aim to criticise the local actors but rather to shed light/give a closer look on the local processes of decision making by AGSOs while dealing with distribution and obtaining of inputs for growing agricultural crops.

government. An example is the state document called 'Shock worker 90 Days of cotton'. This document is issued during the most important period of cotton growing. With the help of this (and other state documents issued in this regard), the regional *hokim* attempts to make sure that everything goes well with regard to cotton and wheat growing. This document states that agricultural service providers amongst other state organisations are responsible for the harvest of 2010 (for details on 'Shock worker 90 Days of cotton' see section 3.4). Another document indicates the importance of cotton for the national economy. This document declares the following regaring a range of state and agriculture related organisations:

> 'This season the Prime - Minister has announced that the working day in agriculture lasts for 24 hours per day!'(Interviews with the officials at F. Co, 2010)

And

> ‚…take into consideration, that leaders of the state agricultural management and agricultural service providers, law enforcement offices, bank servants, the managers of the farms are directly responsible for the fate of cotton…' (The Regulation of hokim of Khorezm Region, Republic of Uzbekistan, dated on 31.05.10, #125K).

One more example of a legal document issued by the Prime Minister of Uzbekistan is a request towards all AGSOs and state administration to respond urgently to bad weather conditions that impact cotton:

> 'Due to the recent rains a number of weeds in the soils increased and a disease as gummos has appeared that caused the 65% of decaying cotton roots. In order to ensure good development of cotton and prevent rotting of the roots of cotton and disease gummosis immediately urgent measures to be taken
>
> 1) Regardless of the type of ownership of the organisation, use existing cultivators (applying the detachment method). One representative from the Organisation of Interior Affairs will supervise every group of workers;
>
> Within two days to draw to the first cotton cultivation the representatives of Ministry of Interior Affairs, specialists from hokimyat, local assets, experienced farmers, scientists, and everyone who has something to do with agriculture;
>
> Fully secure tractors with plows, hoes, harrows and other devices, as well as night lighting.
>
> Ensure a non-stop supply of fuel for tractors on the fields, so they can work day and night;

On daily base, provide the workers of cotton gins and machine operators who take part in the first cultivation with two hot meals daily, payment and necessary working conditions.

2) Application of suspension[21]: ensure follow up of the works,…conduct the initial cultivation, using suspension within three days;

Explain to all farmers about the importance of cultivation and application of suspension;

Urgently provide assistance in application of suspension and provide necessary fertilisers; organize a centralized preparation of solutions in the local fertilisers points, as well as required for that pumps, hoses, people and equipment to work;

Provide in a written form to the Cabinet of Ministers the report on the fulfilled works and held control by the regional hokims, prosecutors, and police department chiefs
(from the Registry in the Fertiliser company, a letter from the Prime Minister Mirziyayev, 2010)

From the legal documents described above, we can see the importance of and details of control over cotton and wheat to the state. AGSOs that play a small role in cotton and wheat production are less regulated and are not held to laws such as the ones described above. The above mentioned state documents also illustrate how centralised Uzbek agriculture functions.

4.2.1 The Machine Tractor Park and the Fertiliser Company in the state procurement system

In this section, I present the responses of the F. Co and the MTP towards assignments related to the state agricultural campaign. Both of these organisations are necessary for the implementation of the state procurement system and respond similarly. However, their treatment by the state differs.

Obeying orders: Frequent and intense interactions with the state

The directors of the MTP and the F. Co, having experienced Soviet times, continue to comply with the state orders under the present day state procurement system. *'The state assignment to the MTP and the F. Co is the first and foremost to be fulfilled'* (interviews, 2010). Participation of the MTP's and F. Co's leaders and staff in the state agricultural campaign is reinforced by the documents like 'Shock worker 90 days of Cotton' (Udarnik 90 days of cotton) as indicated by

21 Suspension – a treatment of cotton to prevent damages by pests

two documents in section 4.2. The readiness of the F. Co's official to participate in the campaign "Shock worker 90 Days of cotton" is shown here:

> For June 2-6: The regional heads of 31 agriculture-related organisation will supervise the condition of the cotton growing and cotton fields (all related to it issues). 10 districts in 5 days we will have to check. We arrive at 06.00 there. The district's territory will be split amongst us. We will be divided in teams of 2-3 persons. For instance, I will be put with land measurer and a representative from the Ministry of Agriculture and Water Resources. We will check the plots (former kolkhoz) on our issues, for instance, I will check in regard supply of fertilisers in that area. But I can check other specifications as well. At 11.00 we will be gathered by the regional hokim and report (interview with the official at the F. Co, 2010)

The directors of both organisations try to carry out orders on time, and they make sure that their subordinates do the same. The resources of both organisations are fully engaged in the control of the agricultural production process during the entire process of cotton growing. To serve state agricultural goals, the F. Co I interned at is assigned a few areas in Khorezm region and usually male staff in the F. Co has to take care of these districts. The assigned representatives are responsible for cotton growing and all relevant activities there. As indicated by the citation below:

> 'I went to my district today, I am assigned as a representative near the Turkmen boarder. Its almost 60 km from here. This morning I have been to the meeting together with the MTPs regarding machinery, there were some problems. Now I go home for lunch. After, I have to go to my district. And after at 18.00 there is a meeting in the hoakimyat regards the cotton seeding and the first agro-chemical measures application progress' (Interview with the official at the F. Co, May, 2010).

To ensure that state orders are fulfilled, one common task of the officials of both the MTP and the F. Co is to help to resolve problems related to fertilisers supply and payment for fertilisers by farmers in the assigned districts. For example, they have to make sure farmers pay for the rendered services. They also have to remind the district bank workers to transfer the money from farmers to the F. Co for the provided fertilisers.

In addition, both directors are mobilized by the *hokimyat* to assist in controlling the harvesting of the cotton fields, and the officials at the F. Co and the MTP supervise farmers in their assigned districts. The former leaders of Soviet *kolkhozes* are the present day directors and managers of AGSOs. Thus, state-ordered-crop production benefits greatly from these trained leaders. The following incident illustrates this point.

'The official from the MTP arrived at the farmer's field. He asked the farmer 'how much cotton have you delivered? - 7 ton; 'it is little. How much are you going to do today? - 2,5-3 tons, and it total it will be 10 tons. The official told the farmer that it was not enough. The farmer: 'what can I do? The cotton is not ready!' The official advised the farmer to cut the roots of the plant. It sounded as unofficial method, so the cotton opens up faster. The farmer was grateful for the advice. We went to the next farmer' (Field diary, 2009).

This phenomenon is a result of the organisational path dependency from the previous Soviet system. The MTP director is an example of a charismatic and energetic former *kolkhoz* leader, with much knowledge on farmers and farmlands. His experience comes from working in agriculture for approximately 40 years[22]. Even though the *kolkhoz* system of supply no longer exists, the *hokimyat* continues using the MTPs' knowledge and localised expertise to serve the needs of the current agricultural system.

The directors of both organisations also have to take part in the meetings convened by the local and regional state administration, superior organisation of AGSO and *selector*. For example, the MTP director participates in approximately 200 meetings each year that are related to the preparation ofand harvesting of cotton and wheat (Interview with the MTP director, 2009). It is one of the many examples where the MTP director and F. Co director together with other former *kolkhoz* co-workers currently still involved in agriculture are mobilised by state officials to accomplish state ordered tasks based on their former, rather than present, roles and knowledge.

Since they are important to the procurement system, the MTP and the F. Co have to be contactable in order to respond to the state demands any time similar to Soviet times. The MTP director always carries his cell phone and can be reached by the state administration representatives day and night. In addition, there is always one person in the MTP on duty until 20.00 hours and over the weekend. In contrast, the F. Co is wealthier and has closer ties to the state administration. Therefore, it has more ways to be reached by other state organisa-

22 Being a part of a big state procurement system, the practice of assigning directors to lead AGSOs exists in Khorezm, Uzbekistan (Trevisani, 2008). During the research, such cases were noticed in the AMTPs and other agricultural service organisations. However, in the case of the MTP this was not the case. Here, the director was not assigned by the state politicians, but by the main shareholder of the MTP in order to be able to restructure the business, rather than to comply just with the state agricultural politics. Therefore, in this respect this case is exceptional and is an example of how difficult it is to restructure the business during the process of agrarian change.

tions (e.g. the *hokimyat* and the apex organisation) than the MTP, including fixed phone, mobile phone, fax, and additional hours of work (night duty). The F. Co employs two operators to receive faxes between 9-18.00 hours and from 18.00-22.00, male staff workers take over the night shift for faxes. There are no exceptions during the holidays. In addition, the F. Co runs a registry of all incoming documents, protocols from meetings, letters, requests, orders, protocols from state-related organisations (e.g. apex organisations, *hokimyat*, court, regional branches). The following field note extract illustrates the readiness of the F. Co director to respond towards the state's assignments at any moment.

> 'I wanted to make an appointment with the director of the F. Co for the interview, but he apologized and said he cannot make plans in advance. As in any moment he can be bothered by the state organisation, apex organisation, local administration and he will have to take care about the given task immediately' (Field diary, 2010).

The example above of how the MTP and the F. Co work signify that activities of both organisations strongly deviate from their official functions. Participation in the agricultural system requires a lot of time from the leaders of both organisations. The functioning of the MTP is negatively affected, because the director, apart from being responsible for all business processes in the entire organisation, is the only person in charge of decisions in the MTP. This is unlike the past, when managers of different departments shared responsibility. However, it is not possible for the MTP to refuse state demands as it may cause problems for the MTP. In contrast, even though the director of the F. Co is also overstressed with a similar workload from the state as the MTP director, the F. Co is much bigger and business does not stop when the director is busy with cotton and wheat production tasks. The MTP also has less support from its apex organisation than the F. Co (see section 4.3), hence the MTP experiences more pressure from the state. For example, the MTP is forced to provide services for free 'for the sake of cotton', which does not occur in the F. Co.

By trying to satisfy the state, the directors of both organisations obtain a good reputation and become an example used by the *hokim* to others. This good relationship with the government is useful to the directors and their organisations, especially during this transitional period, because the directors will be less harassed by state representatives. They may even get some indulgences from the state administration. When he has fulfilled state-ordered tasks, the MTP director is called upon less by the state authorities and can take care of the business. For instance, he can increase profit by renting his machinery to other clients like farmers growing non-state-ordered crops, farmers from other districts and *dekhans* (see Chapter 6 for more details).

Different treatment: Indulgence versus neglect

Although the frequency and intensity of interaction between the directors of each organisation and the state administration in regard to the state procurement system are highly similar, the impact of state demands on the functioning of the MTP and the F. Co differs greatly. Moreover, what each AGSO gets in return for state disturbance is different. The little support that the MTP receives from the state administration (e.g. extension to pay the debts) is incomparable to the solid support the F. Co receives from its apex organisation (see section 4.3.3). The minimal state support is insufficient to solving the MTP's financial problems, the biggest and most urgent issue during the transition process (see section 3.7.2 for the MTP case description).

Changes in state control methods

Last but not least, different methods of power implementation are applied towards the directors of both organisations. The MTP and F. Co directors as well as farmers think that control measures currently are more extreme and do not remember such instances during Soviet times. The harshness of control now, however, is not as effective as before as indicated by the MTP official:

> 'The punishment during Soviet times was without offensive words, it was so strong and it gave effect. Now, it is different. At the beginning it touched us, now we do not care anymore. It does not give such an effect as during the Soviet times' (Field diary, September 2009).

The quotation illustrates the power of the state applied towards actors involved in the state ordered agriculture. It also demonstrates that the power methods employed, which are partially borrowed from the previous system and adapted towards the present day system, do not seem to be effective. Thus, there is a need for new control methods.

4.2.2 A Private Bio-lab 'Aek Durman' and the state procurement system

The private bio-lab 'Aek Durman' is a minor player in the command-and-control structure of the rural economy. The link with the state is far less intensive and it not as exhausting as in the F. Co and the MTP. This is because the bio-lab has a lower direct impact on cotton production within the state procurement system. Thus, compared to the other two organisations, the bio-lab receives less attention from the state. At the same time, they are less hassled by the state with regard to the state-ordered-crops.

For bio-labs that provide some services to the farmers under the state procurement system, a certain level of interaction with the representatives of the state administration is necessary. The directors of other non-private bio-labs participate in state agricultural campaign events organised by the state administration. Events include state gatherings by the Ministry of Agriculture and Water Resources and in *selector* meetings run by the Prime Minister (approximately once per 15 days). Apart from that, state bio-labs are not involved in the additional activities like daily *selector* meetings by the *hokim*, reporting on financial issues of service delivery or supervising farmers' fields. At the other end of the spectrum, the private bio-lab 'Aek Durman' I interned at had barely any interaction with the state in relation to the agricultural campaign. As I show in later chapters, the low intensity link with the state gives bio-labs more space to manoeuvre and develop their business.

4.2.3 Summary

In this section, I illustrated that the F. Co and the MTP allocate a major share of their working time and adjust the daily activities of their organisations to serve state agricultural purposes. Despite this, the F. Co receives much more attention and support from the state in contrast to the MTP. Thus, their responsibilities are the same, but they benefit differently. The benefit in terms of state support during the transition depends on the importance of each AGSO to the state procurement agriculture, the state's interest in the organisation and consequently, the financial standing of each AGSO. The methods that the central government and local administration employ during the agricultural campaign are borrowed from the Soviet command system and seem to be outdated. The state authoritarian regime is revealed by these methods.

4.3 Conclusions

Unequal attention from the state

I conclude that the three studied cases of service providers receive unequal treatment from the state. As presented in this chapter, there is a frequent interaction between the state administration with the F. Co and the MTP in regard to the state procurement system. Similarly, the F. Co and the MTP allocate a major share of their working time and adjust daily activities of their organisations to serve state agricultural purposes. Despite this, the F. Co receives much more attention and support from the state in contrast to the MTP. Thus, responsibilities are the same, but benefit is different. The benefit or state support during the transition depends on the importance of each AGSO to the state procurement agriculture, state's interest in one or another organisation and consequently financial standing of each AGSO. With regards to the studied bio-lab 'Aek Durman', the link with the state is far less intensive and it is not that exhausting as in the cases of the F. Co and the MTP. Bio-labs are less hassled by the state in regard to the state-ordered-crops. Thus, bio-labs receive less attention from the state in contrast to the MTP and the F. Co.

The data presented shows that there are changes taking place in control methods over agricultural service providers for the present day procurement system. State control is a part of the current agricultural system in Uzbekistan and it is a part of agrarian structure's path dependency. However, due to the fact that the present Uzbek government does not provide incentives for work, as it was broadly applied in Soviet times, control has been strengthened as an adjustment to the present day agriculture. As during Soviet times, the state mobilises local state organisations to control agriculture (for more details see Chapter 5). The Uzbek government maintains a tight supervision over the organisations that are closely linked to the state-ordered system of crop production. Thus, regional and district government organisations can be regarded as state's instruments of regulation. To regulate agriculture and utilise the knowledge of former Soviet kolkhoz leaders, the Uzbek state also maintains strong control over AGSOs. In particular, the state mobilises the leaders of AGSOs who are the same people and using the same managing principles that remain similar to Soviet ones (Trevisani 2008). Therefore, for the state, AGSOs are levers of control in the present day agricultural system. Previous studies as well as empirical examples presented in this chapter show how, despite reforms in agriculture, state rule remains dominant in Uzbekistan and the hands of agricultural producers are tied to the state procurement system,. Thus, the agrarian structure's path dependency defines the opportunities and constraints for AGSOs to function within present agricultural system. Therefore, AGSOs are not free to decide how to function and evolve under the present day system. The central government applies a variety of methods to control agricultural producers. In particular, the power methods adapted from Soviet times for communicating with AGSOs employed by the state representatives seem to be inefficient and outdated. Thus, new control methods are needed.

First bricks of a relational typology for AGSOs
The empirical examples in this chapter lay the foundation of the AGSOs' typology during agrarian change in Uzbekistan. Different types of AGSOs have emerged (are constructed) during the transition process, and a comparison of AGSO-state relationships broadens our understanding of AGSOs' function and evolution.

The case of the F. Co represents a *state affiliated AGSO*. It is one that has an intensive relationship with the state organisations in regard to the state procurement system. As a consequence, the responsibilities of the F. Co are modified according to the state agricultural agenda. In contrast, the MTP is an example of a *state neglected AGSO*. It is a provider that interacts with the state administrations in regard to the state procurement system at a similar level as the F. Co. However, although the responsibilities and demands of the MTP and the F. Co are the same, the benefit and state support are different. The F. Co as a *state affiliated AGSO* receives significant support from the central government either via support of the local state administration or via its apex organisation. The *state neglected* MTP does not get nearly as much financial support and back-up from the state, as the F. Co. The MTP is distracted, time-wise and money-wise, by having to interact with many of the state bodies (see Chapter 5). *State affiliated* and *state neglected AGSOs* are tightly bound to the agricultural production goals. Finally, the bio-lab 'Aek Durman' is a case of *state ignored AGSO*s: organisations less important to the state agricultural campaign and of the least financial interest to the state. Thus, bio-labs receive little to no state support.

Expanding responsibilities for some AGSOs, shrinking importance for others

Overall, I found that AGSOs play a crucial role in present day agriculture. The relational typology highlights the importance of the relationship between AGSOs and the government: AGSOs' functioning, existence and survival are strongly bound to the state procurement system and strongly influenced by the central government. AGSOs are not independent producers in the current system. Furthermore, they are not just an interface between the central state and agricultural producers, but also instruments of the government. AGSOs can be regarded as subject, object and means of control in present day agriculture in Uzbekistan.

The current procurement system is path dependent, heavily relying on organisations, actors' agency, and methods borrowed from the preceding Soviet system. By looking at relationships between AGSOs and state organisations, we can see what has been borrowed from the past experiences in order to fill the gaps of and reinforce the present still in formation agricultural system. The material in this chapter suggests that the burden of the dissolution of *kolkhozes* has partly been shifted to AGSOs. The inherited Soviet organisational form of AGSOs defines their behaviour and functioning under the present agricultural system. It is interlinked with the personal path dependencies of the present day AGSO leaders who are the same former *kolkhoz* managers. And they are expected to carry on their roles in order to sustain the present day agriculture. Thus, the present day agriculture builds on the organisational inherited structures and personal structure of former *kolkhoz* leaders.

In the following chapter I will examine the relationships established between the three studied organisations and state elites as well as illuminate the inner dynamics of these organisations. The observations presented in the next chapter provide further support of the typology of service providers during the transition.

CHAPTER 5: PERSONAL AGENDAS AND ORGANISATIONAL MEANS: AGRICULTURAL SERVICE ORGANISATIONS AS PLATFORMS OF ENTREPRENEURSHIP FOR EMPLOYEES AND STATE ELITES

State administration is not judged for what it is
but for what it does. And what it does is measured
in many distinctive ways by different actors (Mendras 2002)

5.1 Introduction

Theoretically this chapter rests on the following concepts: political economy of agricultural production, path dependency, rent seeking by the state organisations. My findings in this chapter contribute further to the discourse on state-periphery conflict. As described in chapter 4, the interactions between AGSOs and the state is rooted in the Soviet planned agriculture and also related to a broader centre-periphery discussion, which gained foothold in pre-Soviet and Soviet Uzbekistan (Ilkhamov 2000). After the end of Soviet Union, the Uzbek President Islam Karimov's politics have weakened the regional power and concentrated power in his hands (Luong 2002). At the same time, the overall Soviet maintenance system has ceased (for more details see Chapter 3). These two conditions have put the still existing regional/local state apparatus into the position where they have to seek for ways to carry on without state support. While Uzbekistan's agricultural sector remains part of a largely untransformed command economy in which cotton and grain are part of a state monopoly, the state employs people working in state administration (local elites and bureaucrats) for purposes of control in agriculture. The inherited internal structure of organisations defines their entrepreneurial strategies – the 'agency' of AGSOs. By investigating relationships between AGSOs and state elites, we delve deeper into path dependency and the behaviour of borrowing from the experiences of the previous agricultural system and organisational structures of AGSOs to fill the gaps of and reinforce the present state procurement system.

This chapter is about AGSOs as a space for local elites' and AGSO employees' personal agendas in servicing their livelihood strategies during the ongoing processes of agrarian change. I discuss how local elites and bureaucrats pursue private material goals by using state political ideology and making use of AGSOs. In particular, I address the following questions: What role do AGSOs play for the state and state elites during the transition period? What role AGSOs do play for their employees during the transition period?

The structure of this chapter is such: the introduction is followed by the section 5.2 where I demonstrate a diversity of the relationships established between three AGSOs and state elites under the present day procurement system in agriculture. Section 5.3 unravels the matter of relationships established between each AGSO and their apex organisations. In section 5.4 I shed light on each AGSO as a space for its employees. In the concluding section 5.5, I summarise the empirical findings and add the new evidence from this chapter to the relational typology of AGSOs.

5.2 Agricultural service organisations as a space for bureaucrats and local elites

In rural society, the authoritarian nature of the government is manifest in the protraction of a command-hierarchy that according to Ilkhamov "dwells rather on the fear of administrative sanctions than on economic incentives" (2000). Direct bureaucratic control, for example, gains predominance, prescribing economic tasks in instructions (Kornai 2008). On a more basic level, a variety of "soft" means of control and intimidation influence the people's behaviour in everyday life and keep them on the track designed by the government officials (Trevisani 2008).

The current state-planned-crop production requires continual interaction between AGSOs involved in the procurement system and the state. Links exist between these AGSOs and the following state organisations: the *hokimyat*[23]; the apex organisation of each AGSO (at the provincial and national level); the tax department; the prosecution office; the *militia*[24]; the Regional Ministry of Water and Agriculture; the state technical supervision service; the state nature protection organisation;the regional statistics bureau; the regional state road police service; *Uzavtodaryatrans*[25]; the Ministry of Finance; Monopoly Committee; and the Regional State Department of Bankruptcy. Apart from their direct responsibilities, they have to make sure that AGSOs and farmers follow state orders under the state procurement system (Ilkhamov 2000; Markowitz 2008;

23 Hokimyat (from Uzbek) - state administration on the local, district and regional levels.

24 Milita (from Russian) – police.

25 Uzavtodaryatrans (from Russian) – Uzbekdarya transport company.

Shtaltovna et al. 2011). In addition to their official functions, these state organisations consequently fulfil the unofficial function of fulfilling control in agriculture. Rather than direct land taxation, the maintaining of a crop procurement system, coupled with the state control of exports, and the regulation of the locally available process (local prices) for cotton, all together create the conditions to maintain agriculture a net contributor to the national economy (Kandiyoti 2003; Trevisani 2008).

With a partial reform of bureaucratic structures and processes after independence, the economic status of state bureaucratic organisations is difficult. In general, the current financing of the state institutions mentioned above is much lower in comparison to what it used to be during the Soviet times. Therefore they have to find ways to improve their financial status. According to (Veldwisch 2006), the official salaries of a *hokim* or any other public position (teacher, nurse, doctor or lecturer) are too low. Thus, one of the ways in which clerks of the still heavy state apparatus make their way through the process of transition is by using AGSOs for personal gains. In practice, this translates into constant stress on AGSOs by state clerks while performing their official role in regard to the state procurement system. As it will be exemplified by the case of the F. Co, in order to manage state checks, the organisation has to pay. Despite the fact it takes place regularly, it showed to be the way many organisation function during the current transition process (Ledeneva 2006). The abuse in the three studied AGSOs varied mainly due to the organisations' different financial status and their proximity to the government.

The district governor (*hokim*) has a key role in the shaping of the implementation of the reforms, and he both holds crucial responsibility and is an important decision maker in agriculture in general a decisive decision maker in agriculture in general (Luong 2002; Trevisani 2008). There are special relations between the *hokim* and the agricultural producers, mentioned in various contexts in different reports and articles (Spoor 2004; Wegerich 2004; Wegerich 2006; Wegerich 2006; Trevisani 2008; Shtaltovna et al. 2011). The *hokim*, who is often called by people the 'landlord of the territory', is responsible to ensure agricultural production. The responses of the current managers show that they see the *hokim* as a trouble shooter for a variety of problems. Hence, there is a certain expectation of the *hokim* (Wegerich 2004: 347). It is expected that the *hokim* will help during difficult periods. It is not possible for agricultural service organisations (Wegerich 2004). It is expected that the *hokim* will help during difficult periods. It is not possible for the AGSO to refuse fulfilling state orders, as it may cause problems later. The firm authority of the *hokim* today leaves little space for manoeuvres for AGSO and farmers.

The state maintains a tight supervision over the organisations that are closely linked to the state-ordered crops production. To regulate agriculture and utilise the knowledge of former Soviet *kolkhoz* leaders, the Uzbek state maintains a strong control also over AGSOs. In particular, the state mobilises the leaders of AGSOs, who are the same leaders in Soviet times using the old Soviet principles. This is another example of agrarian structure path dependency. It shows where the Soviet legacy being re-shaped in a new condition, fitting the new Uzbek politics. Regardless what is written in the status of AGSO, state assigned this unofficial function of control which is valid more than their official functions (Ilkhamov 2000, see chapters 4 and 5). As it will be exemplified by three studied cases, in reality it looks like parallel to providing its direct functions to farmers, i.e. providing fertilisers and machinery, AGSOs are supposed to supervise farmers, provide advice, and provide services for free when necessary.

The MTP directors, bank managers, agencies for the supply of agricultural inputs, water consumer organisations and the district department for agriculture all form the indirect levers through which the hokim can make the farmers compliant to his will. Following the rationale of the current reforms, the farmers can be seen as the "heirs" of the kolkhoz. As such, they also are, as the kolkhoz before and shirkat later used to be, a "building block" of the state system. Therefore, the logic of farming still tries to preserve command system and to enhance the terms of usufruct similar to the command production apparatus (Trevisani 2008).

In this section I will discuss to the extent to which every AGSO offers a space for local elites' personal agendas during the processes of transition. I contribute further to the relational typology by sketching the relationships of each category of provider – state affiliated, state neglected and state ignored –with state elites.

5.2.1 The Fertiliser Company and state elites

The work mentality and activity of F. Co is path dependent and similar to the old Soviet style. The importance of F. Co to the state procurement system involves close cooperation with regional state organisations (as discussed in Chapter 4) and predetermines F. Co's working style. Since this organisation remains closely linked to the state, it continues to rest on the previous working principles, which involved the exchange of favours amongst the state organisations. In the system today, these Soviet rules have been adapted and provide a space for bureaucratic mismanagement. Such behaviour occurs regularly and F. Co has developed ways to accommodate state elites as illustrated by the following examples:

> 'We allocate 2 or 3 million per year for charity. This money includes bribes to the local state organisations' (Personal communication with the employee of the F. Co, May, 2010)

> 'You know, there are representatives of the Prime Minister in each region of Uzbekistan?! So, the Prime Minister's representatives together with the hokim are driving around in jeeps. The local branches of our organisation and farmers have to feed them and satisfy their needs. If someone doesn't respond to their requests, they will be scolded, maybe used as scapegoats during the selector meetings' (Field diary, June, 2010).

Observing employees and the way they deal with such bureaucratic mismanagement showed that such behaviour was not unusual and part of the daily work of the F. Co. Employees do not even evade talking about 'small' cases of mistreatment. The wealthiness of the organisation allows it to deal with the crooked behaviour of state elites.

In the case of the F. Co, its strong financial position is both a blessing and a curse. On one hand, it makes it easier to mobilise government backing in enforcing agreements, on the other hand it attracts various forms of corruption and inefficiency (Van Assche et al. forthcoming). Being so close to the state, it is hard for the F. Co to prevent this abuse. Nevertheless, the F. Co tries. Leadership in the organisation plays an important role (see for more details Chapter 6) and the apex organisation tries to prevent organisational abuse, as it was presented in Chapter 4. This high level of organisational abuse is much higher than in two other case studies as it will be presented in the following.

5.2.2 The Machine Tractor Park and state elites

As part of organisational path dependency, the MTP is strongly embedded in the contemporary version of Soviet rules by the present regime. During Soviet times, when everything belonged to the state and everyone served the state, the MTP director was expected to fulfil any request demanded by or in the name of the state. Today, the importance of the MTP to the state procurement system continues the intensive contact between state bureaucrats and the service provider. As no modern-day rules have been formulated, old Soviet rules are still in use and adapted. The MTP director continues to function as he used to but now fulfilling a much wider range of tasks than officially assigned by the state.

Furthermore, state bureaucrats take advantage of the MTP in other ways. For instance, state officials come to the MTP in the name of the state and use it as a password to obtain services for free. In one case, two men from the state administration came to the MTP and asked for a tractor to plough 0.5 ha. In addition, the MTP's resources are used by bureaucrats for personal reasons. Because they possess the power due to their positions as 'state brokers', government officials have managed to establish non-business relationships with the MTP to fulfil state procurement tasks as well as personal interests. Although the MTP does not have the stable financial standing of the F. Co, it nevertheless continues to serve old roles because:

> 'If we say 'no' to the hokim or someone from above, they will take revenge on us by means of tax inspection or by public prosecution, easily!' (Interview with the senior manager of the MTP, September 2009).

Agrarian change has resulted in the MTP being hardly able to make ends meet. Many things have been stolen and machinery is worn out. Although there is little space for local elites' abuse and the MTP is in a deplorable position, the situation does not stop local officials from abusing the MTP and squeezing all they can. For most transactions, state officials accomplish their goals, whereas the MTP gets distracted, time- and money-wise, by having to interact with many state organisations. For the provision of these services to local elites mandated by its former role, the MTP is paid little or not at all, leading to the organisation's gradual demise. Being politically and financially weak, the MTP offers its services in order to avoid conflict, perpetuating the downward spiral of the organisation. Unofficial use of the MTP's resources by public servants may destroy the organisation. Due to the poor economic conditions of the organisation and the excessive misuse of the MTP's services by state officials, the MTP is not enthusiastic about maintaining its machines in good shape or purchasing new machinery. This negatively impacts on the quality of provided services to other clients and results in poor MTP performance.

The MTP director takes preventive measures and is willing to stop providing services for free to state officials. Instead, he would prefer moving into 'money'-based business versus kinship- and network-based transactions. For instance, he prefers to sell old machinery rather than letting it serve the state bureaucrats for free.

> 'Why don't you buy a machine for defoliation and render this to farmers?" – "It is not profitable. When people in the state organisations find out that MTP has this kind of technique, they will start immediately asking it for them, for their relatives, for other officials and they won't pay for it!'(Interview with the senior staff of the MTP, September 2009)

The MTP director tries to find ways to avoid organisational abuse and attempts to reinvent the MTP. However, it is a difficult task (see Chapter 6).

5.2.3 The Bio-lab and state elites[26]

Having stemmed from the Soviet system, former relations with state organisations (e.g. MAWR and the local administration) drive the interaction of both state and private bio-labs with the state. In contrast to the MTP and the F. Co, bio-labs have hardly anything to offer to the state elites currently. The poor finances and minimal role in the state procurement system has caused bio-labs to be on the verge of bankruptcy. This in turn has led to the situation where relationships between bio-labs and the state are based on friendship, former working relationships, and respect to the higher standing bodies in society, feeling of pity towards their apex organisation (Interview with the director of non-private bio-lab, July 2010). For example, bio-labs provide their services to the state with a different attitude from the MTP and F. Co:

> 'Do you provide bio-methods without payment? - well, if a representative of the Ministry of Agriculture and Water Resources and hokim ask me to provide, for the sake of cotton!' (Interview with the director of 'Aek-Durman' bio-lab, July, 2010)

During the transition process, interactions between bio-labs and the state (via state organisations) have become less frequent. Presently, the state organisations only conduct checks on state bio-labs connected to the state procurement system. These checks do not take place often as compared to the MTP and the F. Co, and the state bio-labs have a relaxed attitude towards the visits from the state.

> 'We don't mind, let them check!' (Interview with the director of the bio-lab, July, 2010)

And

> 'It's over for 5 years! It must be a shame for such a boss, if he appealed to it'! (Interview with the director of the bio-lab, July, 2010)

Furthermore, when the state bodies did carry out checks on the state bio-labs, few cases of bureaucratic mismanagement were reported because of the lack of resources of the bio-labs. These cases were referred to as *'bad habits from the past and a shame'* (Interviews in non-private bio-labs, 2010).

5.2.4 Ifoda: is there a space for local elites?

---

26 Analysis of this section and Section 5.1.1 is based on interviews with the directors and staff of not just the private bio-lab 'Aek Durman' but other private and state bio-labs.

In contrast to non-private bio-labs and the MTP, the regional Ifoda branch supports any requests from the local branches. Ifoda sells products at only one condition – cash. Kinships or other links with local elites or requests to provide products 'for the sake of cotton!' do not work in this case. Thus, it does not have organisational path dependency. Ifoda represents a new type of relationship between agricultural producers and service providers that is based on cash, with no informal relations or kinship involved. There is no intervention from the state and no relation to the state procurement system.

5.2.5 Summary

This section contributes further to the relational typology of AGSOs developed in this thesis. The *state affiliated* and *state neglected* AGSOs, as exemplified by the F. Co and the MTP respectively, are constantly approached by state elites. Interaction with state elites occurs on a regular basis because both are important to the state procurement system. However, the impact of state demands upon the MTP and the F. Co differ greatly. As described before, the F. Co is a financially strong organisation, thus bureaucratic mismanagement does not make a significant impact upon organisation. In the case of the MTP, the state abuse exhausts the MTP's organisational resource and discourages the organisation from providing services. State elites are able to pursue their own goals in both types of organisations.

Private bio-labs and Ifoda, representing *state ignored AGSOs*, have weak links with state organisations and had the least cases of bureaucratic excessive use. As a result, *state ignored AGSOs* can move more rapidly towards the private sector and work in accordance with market principles.

In the next section I will see how the relationships between AGSOs and their apex organisations have evolved through the transition period and consequently has an impact upon AGSO performance in the period of agrarian change.

5.3 Agricultural service providers and their apex organisations

This section sheds light on the relationships established between the agricultural service providers with their apex organisations during the last twenty years of transition. As will be presented, the changes occurred in the apex organisations as well as in service providers and interactions amongst them are shaped by the present day agrarian change. As part of the organisational path dependency, every AGSO has an apex organisation that functions at the district, regional and national levels. Apex organisations used to serve as the intermediate link between the central government and AGSOs (functioning at the local level).

The apex organisations used to support agricultural activities and control fulfilment of agricultural targets in the Soviet system of production. Regional and local branch offices of these organizations were managed from the centre in Tashkent where their head offices were located[27]. Since the system was fully state-managed and the state apparatus was relatively big, there was no space for the private sector.

After the Soviet Union demise, these apex organisations have remained. However, many of them seem to no longer be required in the current agricultural system. Over time, lack of state funding as well as lack of necessity to sustain this heavy state apparatus have led these organisations to undergo changes. Some are dismantled, some of them run into bankruptcy, some become reorganized, merged and remerged with other institutions, in some cases the ministries were closed, recentralized and/or their functions transferred to other institutions(Yalcin and Mollinga 2007). As part of inherited structural path dependency, the apex organisations remain familiar forms of control as during Soviet times, with minor adjustments to the present day agricultural system.

In the rest of this section, I will focus on relationships established between the three studied AGSOs and their apex organisations. The apex organisations are the State Joint Stock Company 'Uzkimyosanoat' (referred as *Uzhimprom* in the text) for the F. Co, the MTP Regional Union for the MTP and the Plant Protection Organisation for the bio-labs. The relationship between AGSOs and their apex organisations is a major part of the link between AGSOs and the central government under agrarian change. This section contributes to the AGSO typology and demonstrates that apex organisations and their daughter organisations receive similar treatment form the state. The apex organisations of AGSOs that are less important to the state are not supported and are in the process of being dismantled in the transition process. In contrast, the more important AGSOs (and their apex organisations) retain a strong connection to the government leading them to be strengthened and develop successfully during the transition process.

27 Present day Fertiliser Company and Fuel supply functions in the same manner as during the Soviet times. Machinery production and supply is an example of organization that has changed through the transition period. The illustration of these service providers can be seen on Figure 6.

5.3.1 Relationships between the bio-lab and the Plant Protection Organisation

The Plant Protection organisation is subordinate to the Ministry of Agriculture and Water Resources (MAWR) and is in charge of all bio- labs. The organisation clings to its Soviet role of control and coordination, even if it is bankrupt and officially has a different role now (Van Assche et al. forthcoming).

The poor status of the Plant Protection Organisation and lack of support from the central government towards the Plant Protection Organisation and bio-labs signifies the low interest of the state to maintain this organisation. The state attitude towards these organisations impacts the relationships between the apex plant protection organisation and bio-labs. Since bio-labs are not tied to the state procurement system (as in the cases of the F. Co and MTP), relations between the Plant Protection Organisation and the bio-lab 'Aek Durman' are inactive.

Unlike Soviet times, when the Plant Protection Organisation was allocating financial and technical support to the bio-labs, there is hardly support from the Plant Protection Organisation to the private bio-labat present. Recently, the only financial support the state bio-lab received from the state was the freezing of debts of the consolidated farmers, which was applied to all AGSOs (interview with the bio-lab director, 2010). In the following ironic statement by the director of the 'Aek Durman' bio-lab, we have a sense of the assistance provided by the state to bio-labs:

> Do you receive any state support to your bio-lab? - Of course! They create me all favourable conditions to work! (The director of the 'Aek Duramn' bio-lab, field notes, July, 2010)

Nevertheless, the Plant Protection organisation tries to sustain as an organisation during this period of change. Since it was established as an apex organisation of the state bio-labs, it primarily targets its services towards its 'daughter organisations'. To maintain relevance and usefulness to the bio-labs, the Plant Protection organisation offers different kind of services to the bio-labs from Soviet times. The services offered include study tours to the capital, study tours to Russia and Germany; provision of minor assistance like advice and inputs for all kind of bio-labs. The Plant Protection organisation also tries to establish new projects, hoping that bio-labs will participate. One such project is a development fund for the bio-labs, intended to encourage them to put money for future benefits. However, before Independence, all services from the Plant Protection organisation towards bio-labs were compulsory, free of charge and covered by the common central system. Now, bio-labs have to pay for the services provided by the apex organisation and they are not compulsory. For example, inputs from the Plant Protection Organisation offers have to be paid for by bio-labs.

> 'We supply them with catchers, sugar and honey. They (local bio-labs) buy it just via us! Why? Because they are not allowed. Well, some may buy it somewhere else' (an interview with the official of regional plant protection organisation, 2010)

As seen from the quote, the products the Plant Protection Organisation offers to bio-labs are nothing unique and they can be purchased from the local market or anywhere else. The price of the input is the same as on the local market and the quality is not very high (field notes, interviews with bio-labs, 2010). In addition, as a part of the organisational path dependency, the Plant Protection Organisation tries to impose its 'assistance, care and control' over bio-labs as it also used to be part of their work in the past. Some people at Plant Protection are not aware of the new status of the labs and follow old directives that are either invalid now, or still valid, but inconsistent with newer aspects of the regulatory framework (Van Assche et al. forthcoming). Therefore, one can say that the relationship of the Plant Protection Organisation and state bio-labs is one of 'imposed cooperation', where the apex organisation tries to sustain itself during agrarian change processes.

Some employees of state bio-labs, either because of a Soviet mentality or out of respect and compassion to the Plant Protection Organisation, follow the apex organisation's directives and purchase inputs from it. However, disappointment occurs from this cooperation. For example, inputs that have been paid for are not delivered on time:

> 'Plant protection owes us money. We transferred them 4 mln. Som[28] for the products (sugar, honey, etc), but they have not delivered it to us yet. We also could buy these products on the market. Whatever is provided by the Plant protection is for money!' (The director of another private bio-lab, interview, July, 2010)

One example was a case involving a tractor where the Plant Protection organisation, looking for financial sources to survive, created an unfair situation for the state bio-labs:

---

28 According to the official exchange rate of the Uzbekistan National Bank on July 2010, 4000000 Som is 1983 Euro

> **A tractor case:** 'District/regional Plant Protection organisation got a loan for 90 mln. Soms at 14% annual interest rate[29]. This sum of money was split between 3 bio-labs that subordinate to plant protection; for that money they bought us/made us buy old machinery, we cannot even use! Now we are struggling to pay off the loan. The money we are paying back to the plant protection, they use 45% for their own purposes!!! We don't use this tractor at all!'

The tractor case demonstrates, on one hand, a high level of respect, reliance, unquestionable fulfilment of tasks/orders of the bio-lab towards its apex organisation. On the other hand, it shows a lack of critical thinking and ability to resist the bio-labs towards their apex organisation.

In this section I looked at relationships established between state and private bio-labs and their apex organisation – the Plant Protection organisation. According to the empirical evidence, this cooperation is imposed by the apex organisations on bio-labs as a result of its miserable status and desire to survive during the transition process. Little support allocated by the present government towards both bio-labs and the Plant Protection organisation signifies that this sector is not a priority of state agricultural goals. As a result, neither plant protection nor bio-labs are actively mobilised for the state procurement system. Since many bio-labs are either private or semi-private organisations and develop on their own, the Plant Protection Organisation is a burden for the bio-labs. The bio-lab alone, being further from the state, providing services to state and non-state-ordered crops, and not having any state backup, can act freely and thus they decide alone where to purchase inputs and do not appeal to the services provided by the Plant Protection Organisation (interview with the director of private bio-lab, 2010).

Therefore, the Plant Protection organisation is the case of an apex organisation that is being dismantled in the transition. Hardly any attempts are made by the state to revive this organisation. It is path dependent, continuing to act the same way as in the past. However, the services it offers are now useless, and therefore this type of organisation is withering away. More evidence of its demise can be seen by recent state action: in 2008, some of the functions of the Plant Protection Organisation were transferred to the fertiliser companies because of poor performance (Van Assche et al. forthcoming).

5.3.2 Relationships between the Machine Tractor Park and MTP Regional Union

---

29 In July 2010 90 million Soms were equal to 44621 Euro according to the National Bank of Uzbekistan currency exchange rate

The financial status and otioseness of the MTP apex organisation (which is the Regional MTP Union) is very similar to the Plant Protection Organisation during this period of agrarian change. The Regional MTP Union used to fully supply MTPs with machinery and spare parts but it stopped between 1991 and 1994 (interviews with the MTP accountant and the MTP economist, 2009). Between 1998 and 2002, the MTP Union provided some administrative assistance to the MTP, like planning certain expenses for MTPs and other financial assistance. In the period between 2003 and 2005, financial assistance from the MTP Union completely ceased. Since then the functions of the MTP Union have decreased to transferring republican and presidential orders as well as decisions from the regional committee and the apex organisation on the national level - Uzbekistan Agricultural machinery (Uzagromashservice) - to the MTPs (see Figure 6). For example, if the MTP needs tyres, it sends a request to MTP Union and the latter in turn submits the request to the republican base of machinery. In Soviet times, the union used to supply the materials to the district MTPs for free; since 2000, the few still available spare parts from the MTP Union had to be paid for (Interview with the workshop manager of the MTP, 2009). Therefore, it makes no difference nowadays whether the MTP buys spare parts from the market or obtains them from the MTP Union. Even though the MTP Union is the parent organisation of the MTP, it no longer offers any assistance or advantages to the MTP because the apex organisation has no funds and is almost non-functioning.

Nevertheless, the regional MTP Union tries to provide services and sustain itself at any price:

> 'We have to render machinery to the farmers even though they do not pay. If we do not give machines, we do not fulfil the plan. Even when we don't work, we still need to pay taxes and salaries. What should we do? That's why we are interested to render services even if farmers delay the payment' (Senior official at the Regional MTP Union, interview, October 6, 2009).

Because of inherited organisational path dependency, the MTP Union continues to consider MTPs as their subordinates, as they used to be during Soviet times. The interviewed senior official talks about fulfilling plans and refers to the activities of the MTP as 'we'. But in reality, the MTP Union has no relation/influence towards the MTP's functioning anymore (interviews, observations, 2009). For the services, the MTP Union provides to the MTP, the MTP has to pay 1% of its annual income (Interview, MTP regional Union, 2009). Paying this tax is a burden for the MTP, since it cannot meet its own ends. This negative view of the Regional MTP Union is indicated by the following:

> 'If the government says we are moving to the market economy, why does it intervene in our work? Why does that regional MTP Union still exist? We feed them, they are spongers. Why do they all [the regional MTP Union] sit in regional administration office and we have to feed them? (Interview, August, 2009)

Thus, the apex organisation is more a burden than of help to the MTP. Over time the link between the MTP and the MTP Union has weakened, especially because the MTP has no need for the Union. The interactions that occur between them are mostly forced by the parent organisation. Its attempts to use its former power are, however, not welcomed by the MTP since the apex organisation provides no assistance and instead is a burden.

During agrarian change, the MTP services are still crucial for the state procurement system but mostly for overseeing state procurement agriculture. Because there are other sources of machinery, the government does not support their position in the economy. Nowadays, the MTP is directly controlled by the local state administration and not via its apex organisation. The central government exercises control over the MTP through the local state administration. Therefore, there is no need to maintain any relationship for the MTP with the MTP Union, and the intermediate apex organisation is no longer needed at this stage. As a result, the MTP Union is gradually being dissolved.

5.3.3 Relationships between the Fertiliser Company and its apex organisation

The relationship between the F. Co and its apex organisation, *Uzbimprom,* differ from the relationships established between the bio-lab and the MTP with their apex organisations. The reason for this is due to the following factors: fertiliser production is a state monopoly industry in Uzbekistan; the fertiliser industry is a financially strong sector of the economy and brings high revenues to the state treasury; fertiliser is a significant input for the state cotton growing. Thus, there is more attention from the central government towards both the F. Co and *Uzbimprom.* This attention is translated into all kind of state support measures towards the F. Co. The following examples illustrate the strong state support towards the F. Co that is provided via its apex organisation: the share in the subsidized loan by the state to the farmers growing state-ordered-crops is 37%, 12,3% for the MTP, and 3,8% for the bio lab (Annex 6). Moreover, the F. Co was given an extension on debt repayment to the factories without having to pay a fine, and 5,44 millions Soms were given by the state to pay the debt to the factories. 5,8 millions of Soms (debt from farmers to the Fertiliser's company) were frozen for 2 years (field diary, 2010)[30], and the state added 30% more funds for the additional purchase of the fertilisers due to the bad weather conditions. In the current system, financial problems related to state-ordered-crops in the F. Co are considered a state-related issue, and state support is strong allowing the organisation to function and maintain itself. As indicated by the F. Co's official:

> 'All resources will be mobilized in order to find a solution to this issue from year to year' (interview with the senior officials in the F. Co, June, 2010).

Therefore, when problems arise for the F. Co, the state administration, the apex organisation of the F. Co, the Ministry of Finance, the state agricultural fund and other state bodies are involved in finding a solution.

The relationship between the F. Co and *Uzhimprom* is driven not just by the central government's agricultural and economic interests but also by the fertiliser factories' interest in profits (see Figure 6). The factories producing fertilisers are under the *Uzhimprom* as well as the F. Co, and fertiliser production and sales are under one roof. Thus, factories keep an eye on the F. Co in order to eliminate the practice of free distribution of fertilisers for state agricultural purposes. This is one more reason for a strong link between the F. Co and its apex organisation. The F. Co is monitored daily by the *Uzhimprom* to ensure provision of fertilisers for state agricultural purposes. Importantly, while checking the activities of the F. Co, the apex organisation simultaneously fulfils state controlling function to agriculture by means of receiving daily reports on application of fertilisers and payment from farmers. In addition, visits to local branches by Uzhimprom take place almost every month.

The F. Co's apex organisation is between two fires: On one hand, it stands for the interests of the factories producing fertilisers and tries to avoid provision of fertilisers without prepayment to farmers. Thus, Uzhimprom, on behalf of the factories, tries to promote business principles and generate profits. On the other hand, it is a state organisation and thus it is supposed to support cotton growing. Uzhimprom is pressured by the state to provide fertilisers to farmers growing state-ordered crops without prepayment. This usually causes financial problems for the F. Co and fertiliser factories. Uzhimprom tries to manage the interests of the state procurement system in agriculture and commercial interests of the factories. However, overseeing agriculture is a heavy burden:

---

30 According to the official exchange rate of the National Bank of Uzbekistan on July 2010 5.44 mln Som is 2697 Euro; 5.8 mln Som is 2876 Euro

> 'We are happy that the Ministry of Agriculture and Water Resources is responsible for distribution of fertilisers, but not our organisation. It is too difficult to carry its responsibility. The responsibility for plant protection was taken from us and thank God for that! If the cotton is in bad shape, there is a huge pressure from the state 'why didn't you do those measures on time, didn't apply fertilisers on time and so on' (Interview, Uzhimprom, 2010).

A result of the conflict between the agricultural goals of the state and profit goals of *Uzhimprom* is that the F. Co is subjected to more and more control mechanisms by its apex organisation, sometimes even directly by the chemical factories. The factories have representatives at the local level to control fertiliser stocks, and they must not be under the influence the F. Co, as illustrated by the following quote:

> 'The factories' representatives are never locals, since that would make them susceptible to pressure from local politics or from the F. Co management that might be influenced by local politics' (interviews with the factories representatives, Urgench, 2010).

More specifically, a representative from the Chirchiq factory works mainly with the F. Co Supply department. He checks the operative reports, payment from districts and puts pressure on the F. Co and the bank workers to pay. A conversation between one such representative and an employee of the F. Co. demonstrates the work scope of the representative:

> 'F. Co employee: so, have you locked the stocks?
>
> The factory representative: I just told to one district yesterday to lock the stock, as they have the fertilisers from our factory. I wanted to close the other stocks, but your boss asked me to wait. I will wait for two-three days, depending on money the F. Co transfers to my factory. If the situation does not change, I will close the stocks in 2-3 days, as I carry personal responsibility for payment from the Khorezm region to our factory (daily observations in the F. Co, June 2010).

Taking into consideration the local resource misuse, *Uzhimprom* representatives to Khorezm admit that state agricultural politics is a burden (interviews with officials at the F. Co, 2010). One can regard the chemical factories then as a force trying to extricate the F. Co from the web of local games, which are skimming of its profits, and from the exigencies of the state procurement system (Van Assche et al. forthcoming). The interests of the fertiliser factories help to drive the F. Co to work according to business principles and prevent it from giving services for free to the state-ordered farmers.

5.3.4 Summary

In summary, the three providers differ in the intensity of their relationship with their apex organisation during the ongoing agrarian change. Factors affecting this relationship include the standing and the role of apex organisations as well as the personal attitudes of employees of the service providers.

The apex organisations of the MTP and bio-labs are not really needed for the present day agriculture, thus they are not supported by the central government. As discussed in sections 5.3.1 and 5.3.2, the apex organisations of MTP and bio-lab are more of a burden than help to their respective daughter organisations. MTP Union and the Plant Protection organisation try to demand the cooperation of the MTP and bio-labs, respectively, and attempt to survive at their expense. Some employees, for example at the bio-labs, continue to comply with directives from the apex organisation because of historical reasons or sympathy.

MTPs are extremely important for the state procurement system but presently they are directly controlled via the local and regional state organisations. Thus, the MTP regional Union is dismantled. Bio-labs (and their apex organisation) are less important for state agriculture and revenue. Thus, they are released from the state attention. The bio-lab (and its apex organisation) illustrates that type of the organisation that will either die with its previous functions (which are not needed anymore) or will have to reinvent itself in this period of agrarian change.

In contrast to the other two types of service providers, the interaction between the F. Co and its apex organisation is very intense. Their relationship is guided by the state's conflicting interests in promoting the agriculture sector and chemical industries: the state tries to provide sufficient fertilisers to farmers while ensuring farmers pay for the supply. For this reason, the F. Co receives much more financial support from the central government than the other two studied providers and is a strong and developing organisation.

With regard to the newly established agricultural service providers such as Ifoda, their parent organisation supports any requests from the local branches of the organisation. As mentioned before, this organisation has no relation to the state procurement system.

In the next section, we will see how the relationships with local elites frame the inner organisational dynamics of each of the three organisations and the consequent impacts upon the performance of AGSOs in the period of agrarian change.

5.4 Agricultural service organisations as a space for employees

Currently, the employment policy of all three organisations is rooted in the Soviet model of labour organisation. This model encompasses a hard-working attitude to work, diligence, to serve the motherland by accomplishing state assignments, and trying to be better than others, which would be appreciated with a certificate or diploma. As indicated in the Encyclopedia of Management, regarding the motivation of labour during Soviet times:

> 'Active participants in the labour process and social activities were noticed by the leaders and moral encouragement were allocated, i.e. certificates of honour, marks of distinction, such workers were awarded with orders and medals. Photos of "shock workers" were hung on the leader board, so that workers know their hero. Additionally to the awarding of a diploma or putting one picture on the wall of honour the financial premium was allocated, the minimum size of which was not less than ten rubles'.

Labour processes and salaries in the MTP are still regulated in accordance with the 'Qualification guidance of employees for office managers and professionals' involved in engineering and economic work in manufacturing plants (State Committee on the Union of Ministers of USSR on the issues of labour and payment 1970). Because there no new labour guidelines have been introduced after the demise of the Soviet Union and many employees are the same workers from those times, the socialist labour principles continue to guide the current agricultural state politics of the state. Together with positive socialist features, the negative features of the Soviet system or the remnants of the Soviet times are still in use as well.

In this section, I discuss the inner dynamics of the three studied organisations as well as the impact of local elites and bureaucrats upon inner organisational functioning during agrarian change. AGSOs and their employees depend on each other to survive. AGSOs are a means of income for their personnel during agrarian change, a period when it is particularly hard to find a job. However, an organisation is able to secure its existence thanks to its personnel. Based on the remnants of the Soviet system, each organisation evolves in its own way.

5.4.1 The Fertiliser Company and its employees

So far, I have demonstrated that the activities of the F. Co are strongly determined by the state and the regular interactions between the F. Co and state elites. The internal organisational structure and dynamics of the F. Co are similarly influenced by these factors.

At the regional level

While the other organisations examined in this thesis can hardly pay their employees (more details later in this chapter), the F. Co employs many for no apparent reason other than handing out salaries. The F. Co has a good financial standing because of strong financial backing from the government. This in turn is due to the F. Co's importance in cotton production, and it is expected by the central government to fulfil all orders in regard to the state procurement system. High pay and good working conditions are offered in the F. Co in contrast to other AGSOs (based on interviews and internships conducted in two other AGSO). For instance, the salary in the F. Co is 400000 Som, whereas in MTP is 120000 Som and in bio-lab it is 70000 Som per month[31].

Table 2: What kind of backup is needed to become an employee of the Fertiliser Company?

Employee 1	his brother works in the headquarters of the Uzhimprom, Tashkent;
Employee 2	he used to be a director of the Urgench district branch, later he moved here (to the regional officce);
Employee 3	he came here from hokimyat; he used to work in the revision department. He knows very well how to write laws;
Employee 4	he works here for 20 or 30 years already;
Employee 5	his mother works in the bank, she is an acquaintance of the director;
Employee 6	his father is the head of the city tax inspection;
Employee 7	his brother works in the city prosecution office;
Employee 8	her father is an acquaintance of the director;
Employee 9	his brother is a big businessman (some people refer him as 'mafia'), also a friend of the director;
Employee 10	his father used to be a big clerk, also came here via the director;
Employee 11	She is there for a long time;
Employee 12	his father was a big clerk and his sister's daughter has married to the hokim's family;
Employee 13	he is here for a long time;
Employee 14	a nephew from regional prosecution office;
Employee 15	per se (on his own).

Source: author's compilation based on interviews, 2010

31 Exchange rate in 2010 1 Euro – 2440 Som, thus 400000 Som – 164 Euro, 120000 – 49 Euro, 70000 Som – 29 Euro.

Such attractive working conditions of the F. Co are also considered as a space for the local elites and bureaucrats. Apart from the misuse of organisational resources by local elites (described earlier in this chapter), many employees in the organisation were recruited to the F. Co due to familial relations to the director or someone in the state administration (see table 2). Approximately one third of the organisation is the employees who got placed in the F. Co without connections to the state authorities. In this way, the personal agendas of state elites affect the internal work environment of the F. Co. As long as the monopoly of the company is maintained and agriculture does not collapse, the flow of money will allow for such inflated pay- lists (Van Assche et al. forthcoming).

The close personal relationships between the F. Co and the local elites and state administration due to the unscrupulous employment methods is reflected in the large number of social events conducted together. As indicated below:

> 'We call people from hokimyat to celebrate with us as we work closely with them' (interviews with the staff at F. Co, 2009)

Representatives of the regional state administration are invited to celebrate together with the F. Co's staff at events such as a manager's birthday, the recent birth of a baby and the weddings of the children of managers[32].

Because employment is based not on merit but on kinship, the F. Co senior official has trouble with the employees. Many, therefore, are indifferent towards goals of the organisation, conflicting or not. Managing them, according to the director, is very difficult (Interview with the senior official at the F. Co, 2010). Most employees are just taking advantage of their positions and do not want to work either according to their work description or to what is assigned in relation to the state procurement system. However, the senior official must ensure that the F. Co's responsibilities regarding the state agricultural campaign are carried out. According to the organisational statute, fulfilment of tasks related to the state procurement system is more important than an employee's job description. One such task of the F. Co (amongst other AGSOs) is to supervise a number of cotton districts during the growing season. The following quotes demonstrate the senior official's attempts to motivate employees:

> 'Representatives of each district have to be in their districts and control everything what's going in there! If somebody does not want to do it or thought that

32 During my 1,5 month internship in F. Co I attended three of such events.

> cannot handle the load during the cotton campaign, write a letter of resignation!' (A meeting in the F. Co, June, 2010)

And

> 'If you do not work during the period of Udarnik 90 days of Cotton campaign [Shock worker 90 days of cotton], I will not look at whose son you are, who your relatives are and or at ones who brought you here, you will be fired!' (A meeting in the F. Co, June, 2010)

Also, at the weekly staff meeting:

> If you saw what the checking lady from the Uzhimprom had written in her report after having checked our organisation, half of you would not be sitting here! But I told her that you are working, you go to the districts and encourage the local branches to on-time payment. I asked her to tell her bosses that I am still coping with you. Do you want me to read a list of people she has suggested to fire?

However, the senior official's threats are ignored. The main reason is that many of the staff are well connected to local elites and bureaucrats (see Box 1). Another reason may be that the assignments they receive related to the state agricultural campaign are not part of their official job assignments, and they do not feel responsible. Many employees are not really bothered by the senior official's anger and it is common to hear:

> 'This kind of hassle is always like that, but they never do a thing' (F. Co employee, June 2010)

At the local level

At the local branches of the F. Co that are spread through the region, a different situation was observed. Dishonest behaviour at this level involves mixing fertilisers with sand (interview with the F. Co employees, 2010; farmers' survey 2009). In the past, giving 10% less fertiliser to farmers than indicated in the contract also used to occur but has stopped in the last couple of years. The fertilisers have to be accompanied by the police and the prosecution officer while transported. Therefore, the mismanagement of fertilisers occurs with the knowledge of local state organisations. This scheme has been in place for years, and the staff of local branches continues to take part in it. Such resource resulted in the conflict of interests of the F. Co goals de jure (that stands for the factories producing fertilisers and expect the income from the enterprise where the local directors are shareholders as well) and local branches making profit on the ground (where the misuse of the resources bring to the lower company income, but feeds the local elites as well as the directors of the local branches).

Similar to the regional office, the senior officials at the local branches of the F. Co are unable to make the employees work and prevent the local branches from corruption. They are interwoven into the local power structures and require help from outside to make changes. One outside source is the apex organisation of the F. Co acting in the interests of the fertiliser factories. To illustrate how it tries to affect changes, I describe a situation where *Uzhimprom*, on behalf of the factories, threatens the regional and local state elites' mismanagement of resources. At the annual shareholders meeting, where the achievements and profits of the company are discussed, I observed the following:

> 'The yearly shareholder meeting was facilitated by the responsible person for distribution of fertilisers for agricultural purposes in Uzbekistan from the central office in Tashkent. He was interested to find out about the weak places of the organisation, about the problems, where the money spent, why the district offices do not accept the recommendations which are given by the conducted external audit and revision has to be sent from Tashkent, why some machinery is bought but it stands broken without making profit to the organisation, why there is so much money spent for unclear reasons instead of benefiting to shareholders' annual dividends? However, he received no answer from the participants, who were mainly the heads of the local branches of F. Co and shareholders at the same time. They looked bored, passive and did not feel involved' (Field diary, the annual shareholders' meeting, June 25, 2010).

Amongst other issues, the representative of the *Uzhimprom* threatened to fire the F. Co employees and asked them to improve their work. The senior official of the F. Co was relieved to have support from the National level. Because the current personnel are all connected through the old system, they are strongly embedded in the long existing working relationships and have relations with local elites. Therefore, it is hardly possible to achieve some positive changes under such circumstances for the senior officials, especially if they are also local and considered an 'insider'. Up to the present, the only way to motivate local employees to work was for the senior official to use threats such as '*Checks are coming*!' However, it appeared that the staff was already used to and indifferent to the threats.

Poor obedience of the staff at both the regional and local levels as well as misuse of resources at the local level illustrate the power of state elites and local politics. The local F. Co branches are strongly interwoven in the nexus of local politics. This union of local F. Co staff with the local bureaucracy makes it a very powerful machine, which is difficult to resist for farmers (who are already suppressed) and for the *Uzhimprom*, whose interests are undermined in such scheme. The F. Co staff at all levels has a poor work attitude that destroys efforts by the senior officials to run the organisation efficiently according to market principles. The local informal schemes that benefit local elites and bureaucrats are entrenched. Local politics and agendas are in conflict with state agricultural goals, and the local branches of the F. Co are not entirely subordinate to the central government. The investigated F. Co plays a role of an interface between two fires, i.e. the *Uzhimprom* (lobbying the interests of the factories producing fertilisers and shareholders' interests) and, on the other hand, it has to cover the drawbacks of its local level organisations, being strongly involved into the local politics/bribery schemes. Therefore, external force is required in order to resolve corresponding problems. However, a few cases where orders to prevent abuse from the National level are issued, they appear to be powerless. Local F. Co branch officers are acutely aware of the contradictory demands placed upon them and answer with flexibility and ambiguity, sometimes with indifference.

5.4.2 The Machine Tractor Park as a space for its employees

The inner organisational spirit and employment of the MTP personnel differs from the case of the F. Co. Current personnel are employed according to the 'Qualification guidance of employees for office managers and professionals involved in engineering and economic work in manufacturing plants (1970)'. MTP uses the qualifications per se as there is no new employment guidance document designed in Uzbekistan since 1970. The MTP tries to be self-sufficient and tries to decrease its costs in order to survive as an organisation during the transition period. The inside organisational dynamics of the MTP differ to the ones in the just described F. Co. Given its poor finances, the MTP tries to reduce a number of redundant workers as the organisation cannot afford having many employees when there is no job for them. In the winter, employees are asked to take unpaid 'vacations' due to the lack of work (see section 7.2.2). Despite the low salaries and irregular payment, the employees prefer to stay here rather than going to Russia. Unlike the F. Co, I found no cases of organisational resource misuse by placing one's protégé in the MTP. By acting in such way, MTP shows private organisation thinking.

By providing employment for its employees through the process of transition, the MTP plays a role of social security net for its personnel. Despite all difficulties brought by the transition process, the MTP collective patiently faces these and works towards a better future. One of the strengths of the organisation is a strong feeling of belonging of its workers. The majority of employees have worked here for more than twenty years, where children follow the footsteps of their parents. The MTP's management and work experience have been passed on from generation to generation. The former Soviet training of the MTP staff has resulted in a relatively successful workplace today: the state-plan is fulfilled on time, production goals are met, a caring and supportive work environment and a high level of discipline. The devotedness of managers and workers to their jobs as well as solidarity of a collective and discipline contribute to the MTP's high level of respect from clients and organisations they work with.

However, since the MTP has insufficient funds, apart from being responsible for all business processes in the entire organisation, the director is the only person who can make decisions. This is in contrast to the past when the director could delegate decision-making to managers of different departments. Managing the MTP, and therewith steering it through the upheavals of the ongoing process of transformation, requires the MTP director to establish strong bonds with diverse state bodies and find ways to improve the financial situation of the organisation (see Chapter 7).

5.4.3 Bio-labs as a space for its employees

Employment in the private bio-lab 'Aek Durman' is completely different from employment in the F. Co and the MTP. During the Soviet times, bio-labs were sufficiently staffed as the payroll was provided by the state. Today, both state and private bio-labs manage their finances very carefully because they have no support. In the private bio-lab, in contrast to the MTP, labour process transformation has happened long ago. As we know from the organisation description (section 3.7.3) bio-labs are very small organisations, usually consisting of one main person – a director- and family members assisting in the lab or hired group of girls to spread the bio-methods in the farmers' fields. Apart from being very careful with expenses and not offering space for state elites to gain personal benefits, the bio-lab provides a caring environment for its employees. Trying to keep production costs low, the bio-lab director tries to run the business alone with the help of family members:

> 'I do everything by myself, try to keep the production costs as low as possible. My two daughters and a grand-son help me in the lab a few hours per day. Daughters come here to clean and work with flour worms. Papers are cut by granddaughters. Two sons work on the land. They grow wheat and barley to produce and feed bio-products. To hire another person, she/he will work for 2 hours and leave; Azad, my grandson, collects bracon[33] and puts it in jars' (Field diary, July, 2010).

---

33 *Bracon* is a simplified version ofHabrobracon,*Juglandis Bracon hebetor(Latin)* – it is a bio-method and it used in bio-labs to infect the grain

Similar observations in regard to the employment were made in four other bio-labs (both private and state). Often, family members are employed since the salary in the bio-lab is between 70-100.000 sums/monthly and it is neither attractive to workers nor possible for the bio-lab to pay many employees. It is also easier pay family members in kind, for instance by buying cloths or topping up mobile phone accounts (field diary, July 2010). Therefore, the bio-lab is of no interest to employee or to bureaucrats.

Also, the leaders of bio-labs (both private and state bio-labs) were passionate about their jobs. They continue to work and strive to develop their business despite their current poor financial status (see Chapter 3 for detailed description).

> 'I am working since 1980s and I have not bought a bicycle since then. One cannot make big money here' (Interview with the director of the investigated bio-lab, July 2010).

Still, a few interviewed bio-labs have expressed a desire in expanding their business, like desire to establish bio-factory, and to provide bio-methods to entire district. This characteristic is a distinctive criterion of the organisation of the state ignored AGSO.

5.5 Conclusion

Personal gains

As the Soviet state support system has ceased, state organisations have become a platform for obtaining private goals from society. In this chapter, I described the relationship between the three types of service providers and state elites and I described the internal work environment of each AGSO. The AGSO is intertwined with the agendas of state elites. I provided evidence that the approach of AGSOs to elites and employees rests on organisational path dependency of the state procurement system. Since the present day agriculture has taken many elements of the preceding agricultural system, supporting state structures (as exemplified by the state organisations and AGSOs) continue to function similar to the previous system. Thus state actors, as exemplified by the state elites, bureaucrats and some apex organisations strongly rely on power from Soviet times in order to fulfil their official assignments and the tasks related to the state procurement system. At the same time, they pursue their personal agendas. To achieve all this, state elites appeal to methods that stem from the Soviet system, including reciprocity between different actors, many informal rules, kinship in order to avoid payment and placement of their protégé in wealthy organisations. For organisations that are important to the state procurement system, such as the F. Co and the MTP, there are more opportunities and more motivations for state elites to misuse resources.

For ongoing conceptual debates this consequently means that not just AGSOs are levers of control in agriculture, as argued in this thesis, but state elites and bureaucrats are levers of the centralised state rule in relation to the state procurement system in particular, that on one hand, serves the state, and on the other hand, abuses it.

Caught in between?

AGSOs that are part of the state procurement system are caught in the web of state-periphery conflict. They both suffer and benefit from their organisational, individual and structural path dependencies. In general, the state and local bureaucrats gain from their interactions with the providers. A few cases where the National level, represented by the apex organisation of the F. Co, tries to communicate and prevent abuse seem to be powerless. Such instance gives us more insight about authoritarian state in the transition environment. And the fight for returns from cotton is being negotiated between the Tashkent and regional/local elites' level and AGSO are out of it. The dynamics and tension between the state and periphery go on.

Standing at the centre of agricultural production, AGSOs cannot only be understood at the 'face value' of their agricultural service provision activities, as 'formally' and 'informally'[34] conducted, but should be understood in addition as having different levels of meaning in their functioning as vehicles for the personal agendas of those employed in and dealing with them. AGSOs are an important resource for employees' and state bureaucrats' personal 'projects' of securing and improving their livelihoods during and due to the ongoing period of change.

I illustrate the organisations' responses in times of ongoing agrarian change, the versatility and creativity of AGSO personnel (including the leaders) as well as state bureaucrats towards AGSO - against all odds. Leadership plays an important role in preventing bureaucratic abuse and sustaining an AGSO through the transition period. This issue is discussed in detail later in this chapter.

Another top-down approach to supervise AGSOs, as presented by the three studied cases, was with the help of apex organisations of the MTP, the F. Co

34 I adopt the distinction of 'formal' versus 'informal' institutions for the purpose of analytic distinction only, while at the same time recognising ongoing debates on the concepts' empirical shortcomings. See (Hodgson 2006; Mielke et al. 2011)

and the bio-lab. The apex organisations were actively used for control in the Soviet times. But now, as it was presented by the cases of the MTP Regional Union and the Plant Protection organisation, they are not as powerful as before, moreover, they undergo changes and some of their functions either transfer to other organisations or disappear as a part of reorganisation process. However, the control over the F. Co and the MTP (and other organisations related to agriculture) carried on by a number of other state organisations on the regional and local level, as it was presented in this chapter. As we could see, the control over the state affiliated and state neglected AGSOs is exercised through the varying relationships established between the MTP and F. Co and the state organisations, and their apex organisations. The use of the control methods in agriculture is one of the facets of the state authoritarian regime in the agrarian change.

Changing importance of apex organisations

In this chapter I also analysed the relationships established between AGSO and their apex organisations. Different relations occur between three investigated AGSO and their apex organisations. The relationships between bio-labs and their apex organisation are imposed by the apex organisations towards bio-labs as a result of its miserable status and desire to survive during the transition process. Little support allocated by the present government towards both, bio-labs and plant protection, signifies that this branch is not a priority for the state agricultural goals. As the result of it, neither plant protection nor bio-labs are actively mobilised for the state procurement system. Since many bio-labs are either private or semi-private organisations and develop on their own, the Plant Protection organisation attested to be a burden for the bio-labs. Bio-lab alone, being further from the state, i.e. fully private, providing services to state and non-state-ordered crops, and not having state backup, can act freely and thus they decide alone where to purchase inputs and do not appeal to services provided by the Plant Protection organisation (Interview, 2010). Therefore, the presented case speaks for one of the organisations that are being dismantled due to the process of transition and there are hardly attempts from the state to revive this organisation. Plant protection organisation is path dependent, as it acts the same way as in the past, however, the services it offers show to be useless at present, therefore this type of organisation withers away. In 2008, some of the functions of plant protection were transferred to the fertiliser companies, because of poor performance (Van Assche et al. forthcoming).

Similarly to bio-labs, the apex organisation of MTP is more a burden than of help to the MTP. Over time the link between the MTP and the MTP Union has been weakening, especially from the side of the MTP. The relationships that occur between them are more pushed by the upper organisations, attempting to use their former power, which is, however, not welcomed by the MTP since the apex organisation does not provide any help, but instead is an additional burden for the MTP. As part of the agrarian change, the MTP services are crucial for the state procurement system. But nowadays the MTP is directly controlled by the local state administration and not via its apex organisation. Therefore, there is no need to maintain any relationship for the MTP with the apex organisation. Thus, the state is not interested to support existence of MTP Union. As a result, the MTP Union is in the process of being dissolved during the transition period.

The interaction between the F. Co and its apex organisation is very intensive. The relationship is guided by the central state's interests in agriculture and chemical industries, trying to prevent supplying of fertilisers to farmers for free. In the view of this, F. Co receives much more financial support from the central government as well as it is strong and developing organisation in contrast to two other studied cases. *Uzhimprom* combines the interests of the state procurement system in agriculture and commercial interests of the factories. One can regard the chemical factories then as a force trying to extricate the F. Co from the web of local games that are skimming off its profits, and from the exigencies of the state procurement system (Van Assche et al. forthcoming). Combining the interests of the factories producing fertilisers and state agricultural purposes stimulate F. Co to prevent giving services for free to the state-ordered farmers. It tells us that there are different interests inside the central state during the transition process.

Relational typology

This chapter further contributes to the typology of AGSOs. The three types of organisations are differentiated by their interactions with state elites, employment strategies as well as the influence from local elites and bureaucracy upon internal organisational dynamics. The more AGSO's business deals with the state (as in the particular case of state procurement system) and the closer the business is to the state, the more instances of bureaucratic abuse are taking place in that AGSO (for more details see section 2.5). *State affiliated* AGSOs, such as the F. Co, are part of the state, and thus having a solid financial resource because they provide important inputs for the state procurement system. This also is accompanied by heavy abuse by local level elites. Similarly, *state neglected* AGSOs are important to the state procurement system and squeezed by state elites during the frequent interactions. In contrast to the other two categories of AGSOs, the *state ignored* AGSOs, have little to do with the state procurement system, are poor and offer little or no space for bureaucratic misuse.

The F. Co and the MTP are constantly bothered by the state elites and where interaction with the state elites occurs on the permanent basis in regard to the state procurement system. The rigorous interaction with the state organisations is shown to have a negative impact upon functioning of the MTP and the F. Co. However, the consequence of the state abuse upon the MTP and the F. Co different. As said before, the F. Co is financially strong organisation, thus bureaucratic abuse does not make a significant impact upon organisation. In the case of the MTP, the state abuse exhausts the MTP's organisational resource and discourages organisation from providing services. Thus, importance of machinery services and fertilisers for the state procurement system indeed sets a frequent interaction between state affiliated AGSO and state neglected AGSO and numerous state organisations. State elites obtain personal goals, as it was illustrated from these relationships. State ignored AGSOs, e.g. private bio-labs, where the links with the state organisation dry out, and a small share of new companies like Ifoda that had no Soviet past, showed to have least cases of bureaucratic abuse. Except the cases of the non-private bio-labs, where the apex organisation makes use of its daughter organisation in order to survive through process of transition. Therefore, private bio-labs and private new established service providers do not provide a space for the state elites to pursue private agendas. Therefore, state ignored AGSO, as suggested by typology and proved by empirical cases from this section, more rapidly move towards private sector and working in accordance with market principles.

The relationships established between the three AGSOs and their apex organisations are another component of the relational typology. The financial and political standing of the apex organisation as well as intensity of interactions between an AGSO and its apex organisation provides additional evidence of the importance of the service provided to the state. *State ignored AGSOs,* such as bio-labs, are organisations that are losing their connection to the apex organisation most rapidly because the apex organisation does not receive any support from the state and is being dismantled. Consequently, bio-labs, being least involved in the state procurement system, are further from state control and quicker to cast away old Soviet rules. In comparison, the F. Co exemplifies the opposite case. It is a *state affiliated AGSO* where the position of its apex organisation, *Uzhimprom,* has been reinforced during the ongoing transition. Also, the link between F. Co and *Uzhimprom* have strengthened due to the importance of revenues from the chemical production to the state and the importance of fertilisers to the state procurement system. Thus, the F. Co and *Uzhimprom* are very close to the state. For *state neglected AGSOs,* their apex organisations have become useless, they receive little state support and are being dismantled during the transition period, similarly to apex organisations of *state ignored AGSOs*. For example, MTP Union is being dismantled and the MTP is directly controlled and mobilised by the Uzbek state via the regional state administrations for the purposes of state procurement system. Weakening of the link between the MTP regional Union and the MTP signifies about MTP's moving status from full state control (as in the case of the state affiliated service providers) towards freely deciding upon whom to provide services and at what price (as in the case of the state ignored service providers).

In the next chapter, I will look at AGSO-farmers relations and draw attention to the response strategies of AGSOs to numerous tensions they face while working under the state procurement system in the period of agrarian change.

CHAPTER 6: SERVICING DE-COLLECTIVISATION: STATE ORDER VERSUS PRIVATE COMMERCIAL PRODUCTION CLIENTELISM

6.1 Introduction

Theoretically this chapter focuses on the agency of former *kolkhoz* leaders, the adaptation of informal rules, and patron client relations in the ongoing period of agrarian change (see section 2.6).

Presently there are, broadly speaking, two categories of farmers (Ilkhamov 2000; Spoor 2004; Trevisani 2008). One category consists of agricultural producers working for the state procurement system; second category comprises farmers and rural households growing commercial crops. Farmers providing for the state procurement system cultivate 100-150 ha of cotton and wheat. These same farmers also grow commercial crops on approximately 10% of their land. A third type of farming is subsistence farming by *dekhan* farmers. They cultivate 0.25 ha mostly for own consumption and for selling at local markets. Subsistence farmers are free from the state system and decide on their own what to grow and what do with their crops (for details see Chapter 3).

There is a great advantage to farmers who grow cotton and wheat under the state wing, as the state provides water and subsidized inputs, and the production system targets exploiting economies of scale. In addition, farmers growing crops under the state procurement system are prioritised by service providers.

> 'Agricultural service organisations provide services to the state crop farmers in the first place even though some other farmers have paid earlier' (Farmers' survey, 2009).

The small share of farmers and rural peasantry growing commercial crops are left alone to get inputs, machinery and so on for their commercial crops. In addition, their main problem is not so much the obstacles imposed by the local bureaucracy, but rather the lack of start-up and working capital (Interviews, 2009-2010). Furthermore, there are restrictions on the size of land for personal ownership (Ilkhamov 2000). Thus, they have a different relationship with service providers than the farmers who are connected to the state procurement system. As reported by Ilkhamov, *dekhan* farms (peasants) enjoy not just official but actual freedom of doing business, particularly in matters concerning the choice of crops and marketing of their products (Ilkhamov 2000). They decide what to grow and how to sell their products. More importantly, they receive payment for their crops in contrast to farmers under the state procurement system (Farmers' survey 2009).

In this chapter I provide insight into the role AGSOs play for their clients during the period of change and contribute further to the relational typology of AGSOs during the transition process.

The structure of the chapter is as follows. Section 6.2 will present the relationships established between the three AGSOs and farmers based on the in-depth investigation of three types of AGSOs. The differences in service provision by the different AGSOs to different types of agricultural producers will be summarised in the concluding section 6.3.

6.2 Is there a difference in providing services to farmers growing state ordered crops and commercial crops?

6.2.1 The Fertiliser Company and farmers

The investigated F. Co only sells fertiliser to farmers involved in the state procurement system. At the same time, it is the only provider to the farmer under the state procurement system to purchase fertilisers using the state subsidised loan. Such status of the F. Co allows it to dictate to farmers how, how much and at what condition fertiliser will be provided. Thus, the relationship between the F. Co and farmers is strongly influenced by the state, as indicated by the quotation below.

> 'How do you check either the farmers are satisfied with the quality of your services and amount of fertilisers? -I think they are satisfied. I didn't hear complaints so far. If there were some, it is very serious here! We would have been punished by the hokim already! Thus, I think the farmers are totally supplied' (Interview with the senior official at F. Co, June 2010).

Thus, if the central state and its representatives on the regional level are satisfied with the supply of fertiliser to farmers, the farmer should be as well. At the same time, bureaucratic mismanagement at the local branches of the F. Co takes advantage of farmers (as discussed in Chapter 5). Ultimately, farmers' concerns with regard to the F. Co, according to the farmers' survey, are hardly addressed by the organisation (see Chapter 3), and the F. Co dictates the relationship with the farmer, where the farmer has little space to manoeuvre.

6.2.2 Machine Tractor Park and farmers

Unlike the F. Co, the MTP has clients from outside the state procurement system. Farmers growing both state-ordered crops and commercial crops carry out business with the MTP. This duality of purpose translates into the establishment of different types of relationships between the MTP and farmers. The main group of MTP clients – and at the same time the major trouble makers – consists of farmers growing state-ordered crops. The MTP, a powerful organisation in the Soviet period, has lost power in the transition period due to: the high level of debt of its main clients, namely farmers and AMTPs; large debts to the tax department and input supply organisations (interviews with the MTP staff, 2009). Either because they do not receive subsidised credits on time or because of their uncommitted attitude to state-planned crop production, farmers hardly ever pay for services provided by the MTP. Moreover, in many cases the MTP is pushed by the local state authorities to render services to these farmers without pay, which results in significant debts—estimated in September 2009 at 253 million Uzbek Soms, of which 110 million was frozen[35] and 143million to be paid[36]. On April 24, 2010, the additionally compiled debt amounted to 19.5million Soms[37]. Growing state-ordered crops is associated with the Soviet, state-planned production system, and hence market mechanisms are pushed out by long-in-use Soviet principles and informal arrangements largely practised by state bodies.

Exceptions to the contract between the MTP and a farmer are due to an unwritten rule. It stipulates that the MTP has to provide services to the state crop-producing farmers as a priority, without queuing or paying for it. Trevisani states that farmers growing state-ordered crops count as a privileged class (2008). Moreover, farmers that occupy 'privileged' positions are given priority to use the tractor first. Farmers or their relatives usually sit on the boards of state organisations (e.g. schools, rural councils, public prosecutors' offices), other agricultural service organisations and have close and well-established contacts with representatives of the local government or used to work in lead-

35 This sum is frozen by the state tax inspection, meaning that it can be paid later and without additional fines or interest payments.

36 According to the official exchange rate on September 15, 2009, 253 mln. Soms was equal to 116 138 Euro, 110 200 000 Soms to 50 587 Euro, and 143mln. Soms to 65 643 Euro.

37 According to the official exchange rate on April 20, 2010, 19469000 Soms equalled 9 250 Euro.

ing positions in the *kolkhozes*. Despite the fact that the current law in Uzbekistan does not allow occupying a position in the public organisation as well as heading a farm at the same time, this situation still occurs in practice. Trevisani's studies showed that in most cases in which an individual is employed as a public officer and also controls and leads a farm, it is because they 'mostly have a base in town: they are businessmen, doctors, *militia* men, teachers, employees of administrative departments or of the *hokimyat*' (Veldwisch 2006; 2008). Thus, they are the first to receive services from the MTP. Furthermore, farmers whose farmlands are located along a main road are similarly in the position to receive MTP services without queuing or paying, as their land will be the first to be observed when state checks are carried out. This causes many of the financial problems outlined above for the MTP. The order in which farmers receive services first was also reaffirmed by a farmer in Gurlen district.

> 'First, the MTP has to provide services to the farmers growing state crops. Before them, only the heads of other state organisations can get services (e.g. gas provision, hokim, director of the school, director of the water management organisation, bank, state inspection). If the head of the MTP refuses to follow the above mentioned order, there is a high chance that he will meet obstacles while receiving services provided by those institutions' (Farmers' survey, Gurlen district, June-July 2009).

In order to avoid running into bankruptcy due to the meagre payments received for delivered services, the MTP director has developed different methods to encourage farmers to pay for the service. Before going to the field, the tractor driver double-checks with the director if this farmer has any outstanding debt. By doing so, the MTP attempts to avoid providing services without payment, or if services cannot be held back due to the farmer's contacts, there at least will be a reminder about payment. Yet, despite outstanding bills, the MTP still has to provide services due to the farmer's importance in cotton production, indicating that informal mechanisms of MTP functioning (providing services first to the farmers growing state-ordered crops) prevails over the formal procedure (providing services on payment basis). Additionally, barter and informal arrangements are accepted due to the poor paying capacity on both sides. Farmers can pay in kind (e.g. grains, livestock or externally purchased spare parts for MTP machinery) and consequently either receive machinery service or are moved along in the queue. These in-kind payments are then passed on to the staff but do not contribute to the reduction of the monetary debts of the MTP and do not show up in the formal accounting system.

In contrast to state crop production, a totally different pattern of relationships has emerged between the MTP and non-state crop-producing farmers. Farmers, after meeting their quotas, can sell their surplus on the local market or bazaars (Oi 1985), which strongly motivates production increases. Before independence, some *kolkhoz* bazaars consequently offered a marketing outlet for smallholders who had no means of transport or access to markets. Similarly also today some local markets serve as outlet possibility for *dekhans* and farmers' crops outside the state ordered procurement system (Wegerich 2006). Trevisani (2008), further supported by Ilkhamov (2002), outlines the free trading of rice on local bazaars in Khorezm, becoming increasingly important for the small producers (largely subsistence oriented) as well as for large farm holders (largely cash crop oriented). The crops of the farmers are sold either at markets outside the province (such as rice) or are sold to a state processing organisation (Wegerich 2006). Similar to payment from farmers growing state crops, payment can also take the form of cash or the exchange of goods and services (due to the lack of cash).

This increased market integration of agricultural small and large-scale producers also improves the MTP's position as service provider. Especially with regard to the production of commercially attractive cash crops (i.e. rice), the MTP workers are ready to work overtime, includingat night, in order to provide the required machinery. Farmers always find money to pay for machine services or inputs required for the production of rice. Consequently, the director strongly encourages his employees to provide these services reliably. Rice as a commercial crop outside of the state order, therefore provides cash income for both farmers and the agricultural service organisations that cater to them.

Farmers, who had an encounter with private service providers indicated that private services are better. For example, unlike with state MTP, a farmer can instruct the private MTP or tractor holder which uncultivated piece of land he would like services for or request for a repeat of the service if it is not satisfactory. If the service is provided by the state organisation, this is more difficult. However, 100% prepayment is required from private organisations. Farmers surveyed indicated that private MTPs are more flexible, the prices are similar to the state service providers or lower, and the variety and quality of products are higher. The farmers also felt that private MTPs try to be more efficient and address clients' needs because they do not have any support from the state (farmers' survey 2009).

6.2.3 Bio-labs and farmers

The private bio-lab was observed to be client-oriented towards all types of farmers. Despite lacking in many areas in terms of service, the bio-lab does its best in order to be able to provide high quality bio-products to farmers. Similarly to the MTP, bio-labs enter into contracts with all who are interested i.e. farmers growing state-ordered- crops, commercial crops and rural households. All clients are treated the same. The director does not have to prioritise farmers growing state ordered crops to farmers growing commercial crops or rural households, unlike at the F. Co and the MTP.

> 'I have eight farmers with whom I have entered into a contract and I will provide them bio-products according to the schedule indicated in the contract' (Interview with the bio-lab director, 2010).

At the bio-lab I interned at, most of the farmers who are clients have known the director for years. These farmers have a great amount of experience in farming and appreciate the good quality products provided by the bio-lab. Some of them even come from other districts especially for his products and services. The director used to be a biology teacher, which is how many people know him. The farmers are his clients and friends. They come to the bio-lab, drink tea, chat and collect the products (jars) they have ordered. Thus, trust and respect have been established between the director (whom farmers call the 'teacher') and farmers.

> 'Do farmers sign that they received jars from you or do you talk about payment when they come? - No, we trust each other. A stranger will not come here' (Interview with the director of the bio-lab, July, 2010)

The majority of farmers do not apply bio-methods (since they are not encouraged by the state, as in the case with all services in relation to the state procurement system) or do not use their services at all in order to reduce production costs of growing cotton.

There are also less motivated clients who come to purchase bio-methods. They are the farmers who grow crops for the procurement system. Either because bio-lab services are not mandatory for state-ordered crops or because money from the state does not cover production costs of farmers, the attitude of these farmers was very different towards bio-lab services. These farmers only come to the bio-lab just if there is an '*alarm*' situation announced at *selector* meetings that pests have been found in the cotton crops and bio-methods have to be applied immediately. Also, according to the contract they enter into with the bio-lab, they are supposed to pick up the bio-product once per week during the season (April 1 - September 1) as indicated on the agreed schedule. However, the farmers often do not follow the schedule or do not pay when they pick up the product.

> 'Farmers say, they do not have money now. One farmer paid 225000 Sums, but took a product for 350000 Sums. I wrote him a letter 'I am very generous. So please transfer money to me when you can' (Interview with the bio-lab director, July 2010).

Even though farmers do not come regularly and do not pay on time, the bio-lab continues to provide services. One reason is due to the specific biological procedures in the bio-lab, the director can only produce a certain amount of product at a time and the product cannot be stored for too long. Next, he does not want to lose a client, even if they realise later that his services are superior.

> 'If I don't give him product, next year he will not come here and will search for other labs. Next year, he will see other labs and will realize that my lab and services are better. And then he will come back. I want to keep my clients' (Interview with the bio-lab director, July 2010).

Provision of services during the agrarian change period requires a lot of patience and creativity of the bio-lab director towards his clients. The bio-lab director tries to find solutions in every case as he needs to secure the income of his organisation by ensuring that clients are satisfied with the service and do not move to other bio-labs. For example, indulgences are applied towards peasants and farmers growing vegetables and melons/water-melons, fruits and vegetables due to the poor paying capacity and the fact that they demand a very small quantity of produce. Thus, they can pay the way they can, for instance, in kind. For instance, peasants can pay by, bringing jars, wheat, and sugar. Often, the price for them will be reduced or the product will be given for free due to personal generosity of the bio-lab director. Rural households also 'pay' by bringing old books, which are needed for the work of the bio-lab.

> 'If a farmer or a peasant requests for a product for 300-500 sums, I give it to him without payment. Anyway, the bio-method will die. This is a very tender work and product. I can't bring it to the market and sell. My main clients are farmers' (Interview with the director of the bio-lab, July 2010).

The studied private bio-lab, in contrast to the two previous cases of the MTP and the F. Co, is an organisation not related to the procurement system. It means that this type of organisation can decide alone what and how to do, and thus, it tries to work according to market principles. The bio-lab, due to its poor economic situation and being rooted in the Soviet system, is unable to provide perfect services. And also it still allows delays in payment for services from the farmers or allows payment in kind, as presented.

In addition to allowing clients to fall behind on payments, lack of knowledge also prevents the bio-lab from improving its financial situation. However, as shown, the director works on both. He tries to improve his service and looks for ways to make farmers pay. Similarly the bio-lab and the MTP try to find innovative solutions how to make clients pay.

6.2.4 Ifoda and farmers

Western companies, as represented by Ifoda, are in the category of state ignored AGSOs, similar to the private bio-lab. Informal arrangements significantly decline when obtaining a service or a product from Western companies. The market principles are dominant in these organisations. Ifoda, which is a representative of the international pesticide company in Uzbekistan and respectively in Urgench, accepts only financial relationships regardless of the client. The relationships that occur between Ifoda and farmers, whether the farmers grow commercial crops or crops for the state procurement system, are very straightforward: the product is sold in return for money.

6.3 Conclusions

In this chapter, I looked at relationships established between AGSOs and farmers. The agency of AGSO leaders and personnel showed to be of crucial importance for the organisations' survival in the face of changes as well as while experimenting new coping and survival strategies during the transition.

State procurement system does not encourage farmers to act according to market principles.

Buying a service or a product for growing a state-ordered crop, for a commercial crop or for personal consumption results in different types of interaction between a farmer and AGSOs. As illustrated, the F. Co is focused on providing services for farmers growing state-ordered crops. The MTP has to supply services to farmers growing state-ordered crops and to a variety of clients. Finally the bio-lab serves all kind of clients. The studied private bio-lab, in contrast to the two cases of the MTP and the F. Co, is an organisation not related to the procurement system. This type of organisation can decide alone what and how to do, and thus, it tries to work according to the market principles. The bio-lab, due to its poor economic situation and being rooted in the Soviet system, is unable to provide perfect services. And also it still has to allow delays in payment for services from the farmers or allow payment in kind, as presented in this chapter. Therefore, traditions and lack of knowledge on how to run a business prevent the organisations from moving forward. However, the director tries to improve his service and looks for way to make farmers pay. Similarly the bio-lab and the MTP try to find innovative solutions to collecting payment from clients. However, in the case of the MTP, the organisation is not afraid to lose a client, but it will be reprimanded by government representatives. In contrast to the MTP, the F. Co does not face such censure and is always supported by either the state administration or its apex organisation.

Western companies, as represented by Ifoda, also provide agricultural services. Informal arrangements significantly decline when obtaining a service or a product from foreign AGSOs like Ifoda.

Adding to the relational typology of AGSOs

As represented by the case of the F. Co, *state affiliated AGSOs* are oriented more towards fulfilling the state orders, thus towards the state rather than farmers. Its close relation to the state (see Chapters 4 and 5) helps to avoid many problems as MTP and bio-lab face with payment from farmers. *State neglected AGSOs* (as presented by the case of the MTP) have multiple clients. It is strongly influenced by the state to provide services primary to the farmers growing state-ordered crops, where farmers have low motivation to pay and because of that MTP faces financial problems. Facing such a situation, the MTP finds innovative solutions to make farmers pay and it is eager to provide services to a variety of clients out of the state procurement system. This serves as an indicator to the MTP's moving status from the state towards market relations or towards category of state ignored AGSOs. And finally *state ignored AGSOs* (as presented by the case of private bio-lab 'Aek Durman') tries to provide good quality of services to all kind of clients, irrespectively of whether it is for state procurement or for commercial farming purposes. Here it is a market driven motive, not the state as in the cases of F. Co and partially MTP, where the motives are a mix of state enforcement and market. *State ignored AGSOs* are those who are eager to move towards market economy thus employ market principles in their work. *State ignored AGSOs* avoid entering into informal relationships with local elites, which are often exercised by AGSOs while working for the state-ordered agricultural production. Informal relationships showed to disturb a lot the work of state affiliated and neglected AGSOs. *State ignored AGSOs* are those organisations that offer better quality of services and missing ones to farmers who are searching for them.

CHAPTER 7: REINVENTION AND DIVERSIFICATION OF AGRICULTURAL SERVICE ORGANISATIONS: THE IMPORTANCE OF LEADERSHIP

7.1 Introduction

As proposed by Long (2001), it is important to characterise different actors and study their motivations to understand the social relations and practices of organisations. Within an AGSO, the leader (director) is a key to the survival of the organisation. The current leaders of the AGSOs examined used to be *kolkhoz* employees or were specialists in their respective fields since Soviet times. Therefore, their expertise in agriculture is especially appreciated by farmers when there are no alternative sources of knowledge and information.

In this chapter, I examine how and to what extent organisational directors and their employees try to reinvent their organisations and sustain them through the process of transition. In the previous empirical chapters, I showed that the financial situation of the organisation is largely affected by the inner and outer relationships of each of the AGSOs, which are summarised in the relational typology chart (see Table 3). The results presented in this chapter contribute to the final components of the relational typology of AGSOs during the transition process.

The structure of the chapter is as follows: based on the in-depth investigation of three types of AGSO, section 7.2 will present a variety of reasons for reinvention of AGSOs. Different motives for diversification of three AGSOs follow in section 7.3. The findings of this chapter are summarised and discussed further in the concluding section 7.4.

7.2 Agricultural service organisations: reinventing or improvising to reduce state abuse?

7.2.1 Is there reinvention in the Fertiliser Company?

The situation of the F. Co is complex. As the state monopoly organisation, it is required to provide fertilisers to farmers growing cotton and wheat under the state plan. This often requires provision of services without payment to farmers, and the F. Co has accumulated a huge debt from farmers. In turn, the F. Co owes payment to the factories producing fertilisers (interviews in the F. Co and with factory representatives, 2010). Consequently, the profit-driven factories put pressure on the F. Co to collect from the farmers. Therefore, the F. Co is caught between two fires, between the central government's interest in cotton growing and *Uzhimprom*'s interests in receiving returns from fertiliser production.

As discussed in chapter 5, the F. Co had to adapt its working processes to serve the needs of the state procurement system. For instance, the F. Co employees have to combine their official job assignment with additional tasks for the state procurement system. Also, being part of the procurement system involves organisational abuse by the state elites (for more details see Chapter 5). Therefore, despite the strong financial position of the F. Co, its senior official tries to undertake some measures in order to reduce the conflicting demands placed on it. For example, in order to prevent provision of services to farmers for free, a new measure is created by the director to show regional state representatives that he is carrying out his job and prevent the F. Co from being blamed for lack of payment. As he states during the weekly staff meeting:

> Giving an assignment to one of the staff members "You write a letter to farmers and let them sign it, saying 'I was warned by the F. Co that I have to collect fertilisers on time and apply it before the indicated period. If I haven't applied fertilisers on time, it is my responsibility". So, every farmer receives this letter and signs it; put all the signed documents from all districts in one folder and it should be on my desk. And when tomorrow during the meeting, when hokim will be saying to me 'you have not provided farmers with fertilisers on time', I will show him the documents signed by farmers! In this case, we will save our skin and our pocket. According to the '90 days of Cotton', everyone is responsible for cotton and thus for the delayed application of fertilisers. Therefore we can be accused for failing to fulfil a plan for the cotton, they can hang the debt on our necks. When we have this document, it will save us. We will explain that we warned the farmers and that they did not use fertiliser on time, so the farmer will be responsible for his deeds '(field notes, June, 2010).

It is a huge burden for the F. Co to cope with the tasks related to the growing of the state-ordered-crops but the F. Co director, being an experienced and devoted leader, cleverly creates a new way to provide fertilisers to farmers who do not pay, while maintaining good relationships with the local state authorities.

In addition, the apex organisation (*Uzhimprom*) of the F. Co also provides support to the F. Co to improve its finances. Since it is part of the Uzbek state and must serve state agricultural goals, *Uzhimprom* cannot completely prevent the F. Co from providing free services to accomplish these state orders. Nevertheless, *Uzhimprom* makes constant attempts to prevent provision of fertilisers to farmers for free and protect the financial interests of the fertiliser factories, which are also subordinate to the *Uzhimprom*. One of the new preventive measures implemented was assigning a factory representative to the F. Co with the aim to control the process of delivering fertilisers to farmers just after payment. As indicated by the following quote:

> 'State dictates us what to do. We have put Shuhrat over there to encourage payment procedure. As he is a factory representative and non-local, the hokim can't put too much pressure on him, as if he were a local, because locals can't say 'no' to hokim' (Personal communication with the official at Uzhimprom, June, 2010).

The factory representative, being an outsider and not linked to local power relations, is there to prevent free provision of fertilisers and bureaucratic mis-management at the regional and local levels. This new method minimises the pressure from the local administration towards F. Co to provide fertilisers without payment to farmers at the local (Khorezmian) level. The effectiveness of the method is described by the representative:

> 'I am acting as a representative from the factories to this organisation since 2004. I can assure that the things are getting better in terms of farmers' payment every year. Apart from that, I am an outsider here. And if I do not agree with some of the hokim's orders I can inform the Uzhimprom' (Personal communication with the factory representative to the Khorezm region, June, 2010).

The factory representative stays in Urgench for months. He usually works with the accounting department, checking the farmers' transfers. He also visits the local branches that distribute fertilisers. He has the power to close the distribution base of fertiliser if he notices that fertiliser is given without being paid for. This is a good example of how the apex organisation assists the regional and local levels of the F. Co to cope with the state abuse of their services.

However, it is still difficult to for the local F. Co branch to resist state orders on the local level in the district and local branches:

> 'We are between two fires - local administration pushes us, telling us to give fertilisers even if farmers do not have money. On the other side, we are squeezed by our above standing organisation (F. Co) forbidding us to give fertilisers without payment' (Personal communication with the director of the local branch of F. Co, June, 2010).

The following is another example of how a director of the local fertiliser's branch, facing state directives to provide fertiliser for free on the local level, still tries to avoid granting fertilisers for free under the pressure of the local state administration:

'Recently I was taken by car by our (district) hokim. He drove me around the farmers' fields, showed the cotton and wheat.

Hokim: 'do you want all of this to be lost?'

Maksud: 'No'.

Hokim: 'then give them fertilisers!'

Maksud 'do you want me to violate the President's decree?[38]'

Hokim 'No'

Maksud 'then we have to make farmers pay/ find a way out'

(Personal communication with the head of the F. Co district branch, June, 2010)

The above quotation illustrates the difficulties in resisting fulfilling orders related to the state procurement system on the local level in particular. It also illustrates that the negotiation processes resulting from policy discrepancies on the national level take place on the local level where the implementation of the different policies reveals their mutual contradictions.

The F. Co is an inherited organisational structure that is being used by the central government for the purposes of the state procurement system. Also, it is used for the purposes of the regional and local state elites, as presented in chapter 5. These two factors help the F. Co to survive financially but, at the same time, prevent it from moving towards market relations. Despite these heavy external and organisational problems, the F. Co still tries to reinvent itself in order to reduce state pressure. However, the reinvention is mostly driven by an outside force that is the apex organisation *Uzhimprom*. Overall, this type of organisation does not change much even by trying to reinvent since it is there to serve the state agricultural campaign.

7.2.2 Reinvention of the Machine Tractor Park

The MTP is also mobilised by the state as part of the state order system, but the dynamics of the MTP's relationship with the state negatively impact the functioning of the MTP, differing greatly from the support that the F. Co receives from the state. The MTP managers and, primarily, the director put great efforts while trying to reinvent and thus to revive the organisation in the current process of agrarian transformation. The MTP has to act or find innovative solutions

38 In 2010 there was a Presidential decree that F. Co amongst other AGSOs provides fertilizers at 100% prepayment.

in the following situations: to prevent bureaucratic abuse of the organisational resource, to find time to for its own business apart from assignments in frame of the state procurement system. Since the director now makes all decisions without the help of managers, he must find these solutions on his own. He is hindered by a lack of knowledge about how to work in the changing environment and also by having to maintain good relationships with the powerful state administration representatives in the present day procurement system. However, he uses his best personal skills and Soviet knowledge to be able to negotiate with the state administration and clients. The critical thinking and good budget management of the director (representing the MTP) is demonstrated by the following examples: he attempted to reduce unnecessary costs by firing redundant workers; he offered services to pay off its debts to other organisation. The success and daily activities of the MTP depend directly and solely on the director and his leadership qualities.

Nevertheless, the continuing transition processes have had a negative impact on the organisation. A number of specialists left and the remaining staff have had to combine their own jobs with the jobs of those who have left to keep the organisation alive. They also have to learn new skills in order to meet the needs

Figure 9: The workers of the MTP. Photo: A. Shtaltovna (2009)

of their clients in the present day agriculture and other sectors. Many employees have not received wages for four up to nine months. In order to deal with these difficulties, the director allows the workers to take on extra jobs (making 2000 Soms per order[39]), in order to gain additional income. The manager of the repair workshop legitimises this practice by stating that *"One cannot be strict with workers; otherwise they will take a run"* (Field diary, August 2009). Despite these attempts to 'outsource' the salaries of his staff, the MTP director had to release ten staff members in 2009. Similar effects were found by Ilkhamov in Ferghana Valley, Uzbekistan where work in former *kolkhozes* often turned out to be unrewarded due to wage arrears for months and sometimes years; 57% of the rural residents had not been paid their wages for 14 months on average (2000).

Apart from the financial and institutional problems of the organisation[40], one of the main challenges for the present MTP management is the lack of experience and training in managing an organisation in a market economy. Management nowadays differs a lot from the Soviet times, during which orders would be issued in a top-down manner, i.e. from the Republican MTP, to the Regional MTP Union, to the District MTP. These orders were little discussed and to a large extent simply fulfilled by subordinate organisations. Today, however, the MTP has to develop its own management plan, taking into consideration its capabilities and indicators from the previous year (plan-actual comparison), and is responsible for generating revenue independently from the state budget. Nevertheless, the MTP director is doing well and is ahead of other AGSO managers and directors. He explains this with reference to his socialisation during Soviet times, when people were encouraged to do things well, and to receive fewer 'reprimands' it was worth doing things well and fast *(Field diary, August 2009)*. Thus, when having fulfilled the state-ordered tasks first, the director of the MTP is less approached by state authorities and can focus on the business, e.g. renting his machinery to other clients (farmers growing non-state-ordered crops, farmers from other districts and peasants). Thus, through the period of agrarian change he has learnt to make money by using his former experience of work during the Soviet times, his intelligence, a broad network of contacts and access to information, all of which are helpful for the MTP's coping with the times of change.

As presented in earlier chapters, apart from being responsible for all business processes in the entire organisation, the director is the only person who can

---

39 According to the official exchange rate on September 15, 2009, 2,000 Soms equalled 0.92 Euro.

40 See detailed case description in chapter 3.

make decisions in the MTP. This stands in contrast to the past, when managers of different departments carried responsibility for their work. Managing the MTP and steering it through the upheavals of the ongoing processes of transformation require the MTP director to establish strong bonds with diverse state bodies and find ways out of the financial deadlock. While good contacts with state representatives are important, the director nevertheless tries to avoid rendering machinery for free. The degree, to which the MTP director aims to keep good relationships with state authorities that influence everyday work activities of the MTP, is illustrated in the following example out of the field diary:

> 'The acting director of the MTP made a phone call. He called the person on the line 'the boss'[41]. The MTP director asks, 'I need machinery for sowing 20 ha of wheat in 2-3 days maximum. You will demand it anyway to fulfil the wheat plan. May I start sowing?' The director was listening for a while and said, 'OK' at the end of the conversation' (Field diary, August 2009).

This is how permission to start ploughing is asked for before the state gives instructions officially and this happens quite often. In this case, his request was denied.

Even with the director's well established contacts with state representatives and his good performance towards state ordered tasks, actual support in solving MTP-management related problems by these state representatives is not guaranteed[42]. The process of finding ways out of the financial deadlock has become

41 While communicating with the state authorities, the director of the MTP usually addresses them with the term 'boss'. Trevisani drew on the parallels between the use of this term and the Uzbek 'yushuli' ('yoshully' is 'yosh ulli', which in khorezmian language means 'high aged', or elder, senior). It is a good example of how pre-existing notions of authority have mingled with new ones, as nowadays in Khorezm the yoshulli is widely understood as the 'boss' in the sense of the person on which one materially depends, to whom one must unreservedly obey. Today, less the village elders, but more the head of the kolhoz or the director of a public administration office, like a bank, the tax inspection, the statistical office, and the like, are addressed as 'yushuli'. See publication by (Trevisani 2008; Shtaltovna et al. 2011)

42 For a similar observation on the relation between *shirkat* managers and *hokims*, see Wegerich (2004: 351) 'Against the general assumption of the strength of affect ties, not all *shirkat* managers having these ties to the *hokim* were able to utilise them during the crisis situation', (Wegerich 2004)

a process of continuous negotiation. In our study area, the data illustrate that it continues to be common for managers in agricultural service organisations to find patrons among local state authorities with whom they enter into a mutually beneficial clan-like relationship. This patron-client network pulls in other influential forces, such as heads of law enforcement agencies, bazarkoms (administrators of local bazaars), as well as representatives of the central government (Ilkhamov 2000; Ilkhamov 2007). A way in which the MTP director tries to make use of such networks is that he tries to find contacts within the tax inspectorate and tries his best to convince the state officer not to withdraw money from the MTP's account.

> 'The MTP director called Erkin (a representative from the tax inspectorate) and asked [begged] him to stop the collection of funds from the MTP's account[43]. So the director asked for the possibility to take 10 million Soms only and leave 5 million for the MTP, as they also have to pay for gas, which otherwise will be disconnected – further impeding their ability to make a profit. He explains to Erkin that the workshop cannot work without gas, and assures that there will be money this month, so that the MTP will be able to pay out the remaining sum at the end of the month' (Field diary, August-September 2009).

Oi (1985) has shown that in developing countries, where formal channels for meaningful participation and interest articulation are weak, individuals regularly pursue their interests through the use of informal networks built upon personal ties. The same phenomenon can be found in Uzbekistan. While talking to a highly-ranked state representative, he tries to show the advantages for both, the 'boss' and his organisation (as well as him personally in the end). By fulfilling the old roles in state-ordered agriculture, the directors of AGSOs assure their position and the potential benefits in patron-client-like relationships with state authorities. The loyalty of the MTP director to the state regularly leads to increased freedom in developing business opportunities for the MTP (apart from servicing the state goals) and facilitates access to state support programs such as purchasing new combines for the MTP at a subsidized rate. Therefore, given scarce resources, these opportunities for the MTP rely on personal indulgences by state representatives to the MTP director.

The director of the MTP, having experienced Soviet times and working now intensively with the state, has decided to obey the state. Thus, he has redefined his position as well as that of the MTP according to the requirements of the

---

43 The money in the MTP's account will be transferred to the tax inspectorate to cover its debts.

current system. He is the first to carry out orders, and he makes sure that those who are subordinate to him do the same. By trying to satisfy the state, he obtains a good reputation and becomes an example used by the *hokim* to others. This kind of good relationship is useful to the director and the organisation, especially during these times of change, because he will be less harassed by the state and might receive some favours in return. For instance, when he has some private or business-related problems, he may appeal to the *hokim*. Similar behaviour is appreciated in the Chinese socialist context, where one of the most important benefits for a team leader is respect and the support that it entails. The client is characteristically the team leader's most enthusiastic supporter and helper and can be counted on to praise the patron's leadership while encouraging others to do likewise (Oi 1985).

Due to the poor economic conditions of the organisation and the excessive misuse of the MTP's services by state officials, the MTP is not enthusiastic about maintaining its machines in good shape or purchasing new machinery, which negatively impacts on the quality of services provided to other clients and results in poor MTP performance. Therefore, the MTP director takes preventive measures to stop providing free services to state officials and instead move into 'money'-based business versus kinship- and network-based transactions. For instance, he prefers to sell old machinery rather than letting it serve the state bureaucrats for free:

> 'Why don't you buy a machine for defoliation and render this to farmers?" – "It is not profitable. When people in the state organisations find out that MTP has this kind of technique, they will start immediately asking it for them, for their relatives, for other officials and they won't pay for it!' (Field diary, September 2009)

The informal relations prevent investment in principle good business opportunities that fit the formal for-profit designation of the MTP.

The director tries to avoid any misuse of his services because such mistreatment of the MTP's resources by public servants may ruin the organisation. Yet, at the same time, reinventing the MTP is a difficult task – caught between the structures, institutions and actors of the past and the so far unknown future. In order to run the MTP and maintain its capacities during the period of agrarian change, the MTP director appeals to his solid Soviet managerial knowledge, his intelligence, communication skills and a broad network of contacts and has access to information and the ability to constantly improvise, applying a 'learning by doing' method.

7.2.3 Bio-lab's reinvention

At the private bio-lab 'Aek Durman', neither corruption nor state misuse of the bio-labs' resources was observed. Therefore, the director is able to focus his efforts on reinventing the bio-lab to keep costs as low as possible. Thus the bio-lab is finding innovative solutions to its production process. In the post-Soviet farming sector, there is a tremendous shift from a high degree of labour division in collective farming to managing farms and providing services as a business. New farmers and AGSOs face great challenges to develop multidimensional knowledge, skills, and experience for running a business (Toleubayev 2009).

To reduce the production costs of the bio-lab, the director must use his creativity and critical thinking. One cost-cutting method is to employ family members and indicate lower salaries in the official financial report, thus so that he will

to pay less tax. For instance, his two daughters come to clean and work with flour worms, while papers are cut byhis granddaughters. Two of his sons work on the plot allocated for the bio-lab needs. They grow wheat and barley to produce and feed bio-products (which are small insects). Azad, his grandson, has worked for two to three hours per day in the last month, and his duties include collecting *bracon* and putting it into jars. In return, the grandson will get a new mobile phone and a computer at the end of the agricultural season. Similar arrangements exist in other bio-labs. The hired workers are relatives or neighbours, as well as all children and grandchildren are involved in the work of the bio-lab. They are usually encouraged to work in return for non-monetary items like clothes, or mobile phone credit, allowing the director is able to reduce production costs.

Another way the director of the bio-lab decreases production costs is by using the services provided by his personal and former professional networks free of charge. Since he used to work as biology teacher at the local school, many of his former students are now his clients, or services providers.

> 'The law says that each farmer needs to have a lawyer. In life things work in a different way. The accountant, who is my former student, helps me; I don't have to pay him. He spends 10-15 min for my work. He offered his help to me by himself, when he got a new PC (the tax payment is done online). I work as a communist. All expenses should be transferred/ put on the society and in return, I work for free!' (Field diary, July 2010)

He also has invented new methods to cool the premises for production to reduce expenses on electricity (which is often centrally switched off and expensive). Despite the simplicity of his methods, they work and overcome the difficulties caused by poor infrastructure:

> 'I have invented a new cooling method a fridge is not reliable due to frequent electricity cut downs. It's a stick with a piece of wet cloth on the edge of it to mop the jars with it. In this way I unscrew from the situation when there is no electricity. Also, it is a good replacement for fluorescent lamp and it burns the insects (Field diary, July 2010).

Figure 10: The director of the bio-lab in his bio-laboratory.
Photo: A. Shtaltovna 2010

7.2.4 Summary

This section has shown how three different organisations reinvent themselves during the transition period. The F. Co is trapped in between the central government's interest in cotton growing and *Uzhimprom*'s interests in receiving returns from fertiliser production. The director of the F. Co therefore started to employ measures so that his organisation is not bothered too much while dealing with the state procurement system. The apex organisation was shown to be a driving force in the F. Co and tries to stop the mismanagement at the regional/local levels. The local inherited structures are very strong and it seemed to be difficult for change to occur if the F. Co is a part of the same structure. An external intervention as was presented by the *Uzhimprom*, seemed to help to introduce changes into the rigid structure. However, it will take a long time to see tremendous differences. Overall, this type of organisation does not change much even by trying to reinvent itself, since it is there to serve state agricultural campaign. The MTP's reinvention was driven mainly by the strong leadership of the director. Reinvention of the MTP included finding innovative solutions in order to prevent bureaucratic abuse of the organisational resource and finding time to do own business apart from assignments in frame of the state procurement system. The success factors for running the MTP include the leadership power of its managers, personal ties of managers, previous knowledge on how to work under the Soviet system and possession of networks, respect to the leader amongst the personnel and vice versa, and possession of a professional farming background. In the bio-lab, which is not profitable currently, the leaders have to focus on cost effectiveness and how they can decrease the production costs. Some cost-saving activities include networking and using kinship, hiring children instead of external labours, inventing new methods for lighting and cooling the premises in the cases of electricity shortage and inability to buy new equipment.

7.3 Diversification and innovative approaches: expanding its activities as an adaptation to the new times or a way to survive?

In this section I will look at diversification activities undertaken by the three service providers. The F. Co launched a new type of fertiliser; it has also started to provide transportation services in the agriculture and public sector. It also has taken over some activities of the Plant protection organisation. The innovative approaches of the MTP are driven by the directors of these organisations to reduce production costs and trying to sustain their organisations through the upheaval of the transition process. The bio-lab does not really diversify but rather tries to strengthen the organisation's activities through the process of transition, as described in section 6.4.3.

7.3.1 The Fertiliser Company's diversification

Because the F. Co has sufficient financial resources, the organisation is able to expand its activities as an adaptation to the current times and to fill the gap on the market of missing services. There are a number of new products and services the F. Co has launched or is planning to launch. One new product is KAC potassium and ammonium nitrate) for farmers outside the state-order system. Seeing a growing interest from the farmers' side, the organisation plans to sell (more of it (interview with the senior officials at the F. Co, June 2010). Also, the F. Co has received permission to start business activities besides providing services solely to farmers under state crops in 2010. It plans to launch a network of shops selling tools for rural households and farmers to generate addi tional income. Due to the good performance of the F. Co, the duties of the Plant Protection organisation, which is in a poor state and located one floor below the F. Co (see section 4.3 for more details), have gradually been shifted to the F. Co.

Figure 11: Poultry production at the F. Co's distribution point. Photo: A. Shtaltovna (June 2010)

'Owing the fact that plant protection is in poor condition, it may be transferred to us over time. We already provide some of their services and plan to diversify our products and activities anyway' (Officials in the Uzhimprom, Tashkent, June 2010).

Another activity that the F. Co has started is provision of transportation services for agricultural and public needs. The F. Co has around 1000 units of machinery, including agricultural machinery. In 2007, AMTPs (see section 3.7.2 for more information) were established at the F. Co's district branches to conduct agro-chemical measures, and the F. Co became an additional source of machinery. This has led to competition with the existing MTPs and AMTPs. Diversification in the F. Co is also state-driven. Due to the fact that there is no city sewage system, the state allocated the task of transportation and spreading of a mix of faeces and manures amongst different state organisation like the Cotton factory, hospitals and other organisations (interview with the head of transport department in the F. Co, June 2010). Thus, after agricultural mechanical works are conducted (i.e. defoliation), the machinery of the F. Co is used to transport compost, manure and faeces to the agricultural fields. It is a diversification of the F. Co activities but imposed by the regional government.

From the F. Co registry:

> 'State order: To Khorezm cotton production, Fuel Supply and Fertiliser Company: to start poultry production at the cotton cleaning gins at the amount of 10.000 heads till June 6; at the fuel depots – start up incubators for chicken; at the fertiliser distribution points – poultry production' (Field diary, 2010)

Thus, diversified activities by the F. Co are based on its solid physical and financial infrastructure and it is driven by both, the F. Co itself and by the state organisations. This case underlines the F. Co proximity to the state.

7.3.2 The Machine Tractor Park's diversification

The MTP has different reasons for and methods of diversification of its activities compared to the F. Co. Diversification activities are targeted at trying to sustain the MTP as an organisation through the process of transition. These activities include allowing workers to do extra work to earn additional income e.g. designing and producing different caps (nozzles) for the foreign machinery. The MTP director appeals to non-financial relationships to settle the financial problems of the MTP. The director through negotiation and bargaining manages to reduce the price while working with banks.

The MTP workers find innovative solutions in times of crisis. For example, the production processes in the repair workshop are slow because of difficulties in obtaining required inputs such as gas and metal. The MTP managers and workers have to buy old or worn out water and gas pipes from farmers. Also, natural gas, which is centrally provided by the state, was cut off because of the MTP's debt. Hence, workers buy it from neighbouring organisations, transporting it in combine-harvester tyres. The prices of spare parts vary on a daily basis, which

Figure 12: MTP worker uses old pieces of steel to produce missing spare parts. Photo: A. Shtaltovna (2009)

again makes it difficult for the MTP to budget for parts that are required.

The transition process and financial problems of the MTP push the employees to be innovative. For instance, they have designed and produced different caps for foreign machinery, and then adjusted accordingly their ploughs, cardans, clutches, etc. Such modifications solve the problem of getting expensive spare parts for foreign machinery, and with these nozzles the machinery can fulfil extra operations, which in turn results in extra profit possibilities for the MTP. As the manager of the repair workshop says: *'This is a way out of the deadlock'* (Interview, October 2, 2009).

While in the past all inputs were supplied by the state, and the mutual settlement of accounts (if such occurred) was easily solved with the help of the state, now the MTP has to achieve financial independence. Moreover, the legal status of some providers has changed, formally reducing state interventions into everyday business. Informally, nevertheless these state interventions continue to be part of everyday work. Just like the MTP, many of the state input-providing organisations are in poor condition and face financial problems due to the transition from the state supply system to the private management of their businesses. For instance, the state electricity company and a repair factory are in a bad economic situation, as the organisations they provide services to owe them money. To remedy the deplorable financial status of the MTP and input-providing organisations, the MTP director mobilises his personal agency, kinship and other networks to negotiate the extension of arrears terms. For example:

> 'The representatives of the repair factory, while driving around and checking the factory's debtors in Urgench, arrived at the MTP and talked to the director about paying back the debt. The director asked them to wait for another 30-40 days, when the MTP would have sold old machinery and be able to pay. The MTP director: 'We also have a huge debt to the tax inspection (40 million). First we will sell machinery, after this the (administrative) building and then we can pay out the bills' (Daily observation, September 2009).

The MTP director appeals to non-financial relationships to settle the financial problems of the process of redefining his MTP. For instance, he suggested to a water supply company that the MTP repairs its equipment and writes off the outstanding water bill. Both sides are interested in this operation (field diary, September 2009). Informal arrangements significantly decline with the emergence of market principles while working with the representatives of Western machinery providers, i.e. New Holland and Claas. When working with them, the MTP director cannot mobilise kinship and networks. Or while purchasing spare parts on the open-air markets or from the merchants from Turkmenistan who drop by at the MTP, the director through negotiation and bargaining manages to reduce the price. Nevertheless, the deal will not work without cash.

When working with banks, the MTP director uses all kinds of negotiations to reach his goals. Most of the banks that provide services to agricultural actors (e.g. AGSO and farmers) are under state control. However, establishing relationships with bank employees by giving them presents can improve the situation and quicken transactions. Thus, relationships with organisations or individuals that are not state agents require financial emoluments, otherwise they do not cooperate. In situations like these, with no cash in the organisation, the director will attempt to instigate informal arrangements or use his negotiation skills or some of his former links in order to find a solution for his organisation and workers. Thus, informal relations compensate institutional shortcomings of the Soviet past as well as the state-ordered system of crop production. Consequently, the spaces for development according to market principles remain limited. However there are signs of development according to market principles presented by the examples of the trading of rice and of the trading of spare parts for agricultural machinery that were discussed above.

Despite the current problems, the MTP is looking towards the future. The MTP director intends to strengthen two main business processes and to diversify service provision, by paying off held back wages to the workers, debts on taxes and a bank loan, selling an MTP administrative building, subletting parts of the MTP's land and selling old machinery. As such the MTP is mobilising its long-term capital to cope on a short-term basis with the impacts of the ongoing processes of agrarian transformation and change.

In summary, in order to survive through the process of transition, three organisations, due to their different economic and political standing, have diversified their activities using different rationales. Due to high financial turnover and state's favouring of the F. Co, it introduces a new variety of products in order to secure its position on the market. It also takes on some duties of weaker AGSOs like the Plant Protection organisation and the MTPs. A state neglected AGSO, such as the MTP, reinvents itself to fulfil its state obligations while creating new business activities in order to survive as an organisation during the current transition period. The MTP has shown the best leadership skills and agency, critical thinking, a very considered way of spending money, firing redundant workers, offering its services in return for its debts to the partner organisation, allowing workers to do extra works to earn additional income, and an innovative approach to problems.

7.4 Conclusions

This chapter analysed the reinvention strategies undertaken by three different types of AGSOs in external and internal organisational dimensions. In particular, I analysed strategies developed by AGSOs to prevent misuse of the organisation's resources by the state for the procurement system and to diversify activities during the ongoing period of change.

The final component of the relational typology of AGSOs

This chapter completes the relational typology of AGSOs within the scope of this thesis. This final part of the typology captures the qualitative differences amongst the three service providers with regard to their attempts to reinvent themselves and diversify their portfolio. *State affiliated AGSOs* are more oriented towards fulfilling the state orders. Thus, these organisations prioritise the needs of the state over that of farmers. Also, the close relation of these AGSOs to the state helps to avoid many financial problems like those faced by the MTP and bio-lab e.g. delayed or no payment from farmers. *State neglected AGSOs* (as presented by the case of the MTP) have multiple clients. It is strongly influenced by the state to provide services primary to the farmers growing state-ordered crops, where farmers have low motivation to pay and because of that MTP faces financial problems. Facing such a dilemma, the MTP finds innovative solutions to make farmers pay and is motivated to provide services to clients outside the state procurement system. This serves as an indicator of the MTP's shifting attention from the state towards market relations. If it is unable to succeed in financing itself, it will likely become a state ignored AGSO. And finally *state ignored AGSOs* (as represented by the case of the private bio-lab 'Aek Durman') try to provide good quality services to all clients, irrespective of whether it is for state procurement or commercial farming purposes. Here the motive is entirely market driven, with no impact from the state. In contrast, the F. Co and the MTP motives are driven by both the state and market principles. State ignored AGSOs are eager to move towards a market economy, employing market principles in their work instead of entering into informal relationships with local elites, which is often the norm for AGSOs working for the state-ordered agricultural production. Furthermore, state ignored AGSOs, unlike the other two types of AGSOs, offer better quality of services and are eager to provide new services to farmers who are searching for them.

As I have argued, state-affiliated AGSOs show lower ability to resist state ordered and tasks from the bureaucrats (which is strongly embedded into the system of relations from the past). Thus reinvention in the F. Co was seen in regard to reduce the influence from the state administrate upon the F. Co while mobilising F. Co for state cotton campaign. In diversification of organisational activities, the apex organisation showed to be a driving force. In contrast to state neglected and state ignored (MTP and bio-lab respectively), where reinvention is a struggle to survive and moved by agency and leadership of the organisational leaders and personnel trained in Soviet times. Thus, in state affiliated AGSOs if changes occur, they are state driven. In state neglected and state ignored AGSOs that are being released from the state, change is driven more by the leaders of the organisations. In addition, transition towards market economy from the state-planned economy for both types of AGSOs is faster as compared to state affiliated providers. The state affiliated AGSO remains under the state's wing, thus more state-driven and continue to fulfil state control functions in agriculture.

Survival of an AGSO heavily depends on good leadership

One factor influencing the success of an agricultural service provider during the agrarian change process as in Uzbekistan include the leadership power of managers, personal ties of managers, previous knowledge how to work under the Soviet system and possession of networks, respect to the leader amongst the personnel and vice versa, and possession of a professional farming background. It is a similar finding as in the studies of Uzbekistan transition by Markowitz (2008), Trevisani (2008) and Ukraine by Shtaltovna (2007). As noticed by Ledeneva, regarding people in Russia during the transition period: 'Our people are so inventive, they immediately start finding ways around everything' (2006). In the particular context of Uzbekistan, where the broader structure is characterised by an authoritarian regime and strong state influence upon many processes in agriculture and all actors related to agriculture, agency of actors is important but not always enough to sustain an enterprise in the transition period. One needs solid political ties and networks, as well as political backup as it was in the case of the F. Co or state affiliated AGSOs. For organisations, such as the MTP, that do not much have political support, leaders are likely to become frustrated after having tried hard and give up any attempts to reinvent the organisation.

CHAPTER 8: RESEARCH FINDINIGS AND POLICY RECOMMENDATIONS

8.1 Introduction

The questions my thesis answers were: how have social relations of production between the state and agricultural producers as exemplified by agricultural service organisations changed in the context of transition? More specifically, how have agricultural service organisations evolve and what is the role they play during the process of agrarian change in Uzbekistan? Theoretically I approached the study by looking at agricultural service providers as an interface between the state and farmers. The suggested structure between the main actors involved in agricultural production is that of 'state – AGSOs – farmers'. I have investigated the relationships amongst the state, farmers and agricultural service providers as well as the role of each actor in this chain in the context of agrarian change. To answer the main research questions, the thesis undertook an in-depth analysis of the functioning of agricultural service providers during the transition process in Khorezm province of Uzbekistan. For the present analysis, the thesis' findings will be summarised based on the empirical findings and research questions in section 8.2. Section 8.3 discusses the methods employed in this thesis. Section 8.3 reflects upon the heuristic tool, the 'relational typology of agricultural service organisations' that was developed and mobilised to further understanding of interactions and interdependencies amongst the actors in the ongoing transition process in Uzbekistan. The relational typology distinguishes between different types of agricultural service providers that have emerged during the ongoing agrarian change in Uzbekistan. Theoretical findings are presented in section 8.4. In particular, I discuss what my thesis contributes to our understanding of agrarian change in Uzbekistan and the nature of an authoritarian state. The chapter closes with section 8.5 in which I suggest policy recommendations for the further development of agricultural service providers in Uzbekistan.

8.2 Empirical findings

In the following I summarise the thesis' empirical findings.

(1) Centralised state rule remains dominant in Uzbekistan, and the hands of agricultural producers are tied to the state procurement system despite the individualization of farmers and claims that the country is moving towards a market economy. Framed by ongoing agrarian change and the state procurement sys-

tem, many service providers and most state organisations at the regional and district levels can be regarded as state instruments of regulation and incentive structure creation intended to facilitate and enhance agricultural production activities.

The nature of economic transition in Uzbekistan can be described essentially as a deliberate attempt to maintain the concentration of all economic control and activity in the central government, which directly supervises a gradual and carefully calculated transition to market economy (Luong 2002). While in Soviet times the communist ideology supported the state call to join in the cotton harvest, today this missing ideological embedding is compensated for only by the exercising of state control in agriculture (see Chapter 4). To fulfil the agricultural plan, the Uzbek government needs 'an army' of people who are aware of the preceding, Soviet agricultural system and are able to control agricultural producers. For that reason, the central government heavily relies on the past experience of individuals and organisations of the Soviet agricultural structure, perpetuating path dependency (see section 2.4). The former leaders of the kolkhoz are today's directors and managers of the agricultural service organisations. The directors of the MTP and the F. Co, having experienced Soviet times and working so intensively with the state under the present procurement system, continue to comply with state orders. The assignments to the MTP and the F. Co with regard to the state procurement system are the first and foremost to be fulfilled (field notes, 2010). The participation of service providers' leaders and staff in the state agricultural campaign is reinforced by documents such as 'Shock worker 90 days of Cotton' (Udarnik 90 days of cotton) as discussed in Chapter 4. The resources of the MTP and the F. Co are fully engaged in controlling agricultural production during the entire process of cotton growing. Following the Soviet tradition of staying in touch with the centre, the MTP and the F. Co are constantly in touch with the state in order to be able to respond to the state demands. Participation in agricultural campaigns demands substantial working time from the leaders of both organisations.

The F. Co director tries to carry out orders on time and tries to make sure that those who subordinate to him do the same. While serving state agricultural goals, the following is required from the F. Co and its staff: the company is assigned a few areas in Khorezm region and usually male employees have to take care of these districts. The assigned F. Co representatives are responsible for cotton growing and all relevant activities related to it. They have to help to resolve problems related to fertiliser supply and payments for fertilisers by farmers in the assigned districts. In addition, they have to contact farmers to make them pay for the rendered services. They have to remind the district bank workers to transfer the money from farmers to the F. Co for the provided fertilisers.

Similarly to the F. Co, the MTP director is regularly mobilized by the regional state administration for purposes of the state procurement system. The director of the MTP has worked in agriculture for approximately 40 years. Therefore and due to the MTP's formerly powerful role in distributing machinery, running the repair station and supplying spare parts to collective farms, he is a carrier of knowledge on farmers and farmlands. Even though the kolkhoz system of supply no longer exists, the hokimyat continues to use the MTPs' knowledge and localised expertise to serve the needs of the current agricultural system (Shtaltovna et al. 2011). The directors of both organisations have to take part in the meetings convened by the local and regional state administration, superior organisation of service providers, the selector, etc. For instance, the MTP director participates in approximately 200 meetings each year related to the preparation of, and actual harvesting of cotton and wheat (Interview with the MTP director, 2009). It is one of the many examples where the MTP director and F. Co director together with other former kolkhoz co-workers, currently still involved in agriculture, are mobilised by state officials to fulfil state ordered tasks based on their former, rather than present roles and knowledge. Thus, the functioning of the F. Co and the MTP are fully shaped by the procurement system in agriculture.

The central state continues to heavily rely on Soviet methods of managing collective agriculture in dealing with present day agriculture. This includes the plans for cotton growing, issuing legal acts and documents in order to draw the attention of the F. Co, the MTP and other state organisations to the primary importance of the state procurement system allocating subsidies, involving the state administration and service providers for the exercising of state control over agriculture (see section 4.2). In order to ensure that the central government's requests in relation to the state agricultural campaign are fulfilled by the majority of service providers, control measures by the state organisation are implemented. Amongst control methods the Uzbek government conducts daily meetings, visits to the organisations, conducting supervisions of the farmers' fields, the selector meetings and others. Control over farmers, the studied MTP and F. Co is exercised via the local government (see Chapter 4).

Not just some service providers are levers of control in agriculture, as argued in this thesis, but state elites and bureaucrats are also levers of control in the centralised state rule of Uzbekistan (Chapter 5). To manage farmers and service providers, a number of state organisations are involved. These are departments of the local, district, and regional state administration, police, a prosecution office, a tax inspection, a state technical supervision, a mechanical inspection and other state organisations. These organisations play the role of 'state brokers' (Powell 1970). The state organisations, apart from their official duties, are involved in the control process in the state procurement system. Local elites and bureaucrats pursue private goals parallel to their official functions and functions assigned to them with regard to the state procurement system during the ongoing processes of agrarian change.

The established relationship between the studied cases of the MTP and the F. Co and state organisations indicate that service providers are not independent actors in the current system, despite the fact that the system is proclaimed to be 'moving towards market economy'(Karimov 1995; Wegerich 2005; Trevisani 2008). An important part and the specificity of the Uzbekistan situation of the existence of many service providers and other state-ruled organisations is the continuation of state control exercised in agriculture. Thus, such service providers as the MTP and the F. Co can be regarded levers of control and their practices are both economic and political in the present day transition process.

(2) Agricultural service organisations provide a basis for obtaining private and collective goods for their employees and for other state actors involved in state agricultural production during the transition process in Uzbekistan.

In contrast to the financial system during the Soviet time, the Uzbek government has cut finances toward the public sector. Nevertheless, a substantial number of state organisations have lacked government support since this transition period (Chapter 5). Therefore, the present day employees of the state apparatus, while fulfilling control functions for the state procurement system, appeal to different ways (mainly relying on the previous power of their organisations and informal rules) in order to survive as individuals and as organisations during the transition period. In addition, agricultural service organisations provide means for the survival of their personnel and apex organisations (as in the cases of the MTP and the non-private bio-labs) during times of agrarian change, times when it is particularly hard to find a job. There is an impact from local elites and the state bureaucracy upon inner organisational dynamics of service providers by working towards the employment of their protégés by wealthy service providers such as the F. Co.

Mostly state affiliated and state neglected service providers (as exemplified by the cases of the F. Co and the MTP) showed to provide a space for pursuing the individual goals of state elites and bureaucrats (Chapter 5). One of the ways in which clerks of the large state apparatus make their way through the process of transition is by abusing service providers for personal gains. The F. Co and the MTP are constantly bothered by the state elites on a permanent basis in regard to the state procurement system. The rigorous interaction with the state organisations showed to have a negative impact upon the functioning of the MTP and the F. Co (Chapters 4, 5 and 7).

According to the suggested typology (sections 2.7 and 8.5) the F. Co's close standing with the central government's interest as well as its service is vital for the state procurement system. These two conditions predetermine a close and intensive cooperation with the state elites and bureaucrats on the regional level. Since this organisation remains closely linked to the government, it continues to rest on the previous Soviet working principles and it involves the exchange of favours amongst the state organisations. The abuse cases from the side of state institutions towards F. Co take place regularly and are referred to as usual organisational matters by the F. Co's employees. Apart from the local elites' abuse of the organisational resources, many employees in the organisation got recruited in the F. Co thanks to the kinship of their family members with the director or members of the state administration. High salaries and good working conditions are offered in the F. Co in contrast to other investigated service providers (based on interviews and internships conducted in the bio-lab and the MTP). Such attractive working conditions of the F. Co are also considered as a space for local elites and bureaucrats. Approximately two thirds of employees of the F. Co maintain close connections to the state authorities. Another indicator of the close interaction between the F. Co with the local elites and state administration is a substantial number of social events conducted together. Those can be dedicated to the F. Co managers' birthdays, newborn babies received in the family, marriages of the sons/daughters of the managers of the F. Co, where the members of the state organisations participate and socially bond. The F. Co senior official has trouble with the employees, who were employed via kinship relations with the state bureaucrats. The senior official tries different ways how to make employees work. Thus, the intense relationships between the state elites and the F. Co impact the functioning of the F. Co.

The MTP provides a space for local elites' abuse during and due to the transition period. Occupying state positions and partially fulfilling a state control function, by 'state brokers' has resulted in establishing non-business-related relationships with the MTP, which equates to an abusive use of the MTP services. Even such a deplorable position of the MTP does not stop local officials from abusing it and squeezing all out of it. In most transactions, the state organisation get what they want, whereas the MTP gets distracted, time- and money-wise, by having to interact with many state organisations. For the provision of these services to local elites mandated by its former role, the MTP is not or to a limited degree paid, which results in the organisation's gradual demise. Being politically and financially weak, the MTP offers its services in order to avoid conflict. In addition, the MTP director is expected to fulfil any request demanded by or in the name of the state as during Soviet times, when everything belonged to the state and everyone served the state.

The inside organisational dynamics of the MTP differ from the ones in the just described F. Co. Given the financial deadlock, the MTP tries to reduce a number of redundant workers as the organisation cannot afford having many employees when there is no job for them. Moreover, the remaining personnel have to combine their own functions with the functions of the employees who have left organisations. In addition, they have to learn new skills in order to meet the needs of their clients in the present day agriculture and outside the agricultural sector. In the MTP there was no cases found of organisational resource abuse by placing one's protégé in the organisation as it was discussed in the case of F. Co. By providing employment for its employees through the process of transition, MTP plays a role of social security net for its personnel.

The consequence of the state abuse upon the MTP and the F. Co differ. As said before, the F. Co is a financially strong organisation, thus bureaucratic abuse does not make a significant impact upon organisation. Due to the poor economic conditions of the organisation and the excessive misuse of the MTP's services by state officials, the MTP is not enthusiastic about maintaining its machines in good shape or purchasing new machinery. This negatively impacts on the quality of provided services to other clients and results in poor MTP performance.

In contrast to the two above discussed service providers, a bio-lab 'Aek Durman' has hardly something to offer to the state elites. According to the relational typology (section 8.3) and the case description of the bio-lab 'Aek Durman' (section 3.7.3), all bio-labs are on the verge of bankruptcy. Thus, in the situations when the state bodies come to carry on checks to the bio-labs there were little or no situations with abuse reported. The bio-lab does not provide any space for the state elites' protégé during the transition, except the family members of the bio-lab owner. Bio-lab, having a poor financial status, tries to develop its business, and for that it showed to be very careful with spending money and employing extra labour.

Service providers as a space for apex organisations (section 5.3): As a matter of time and little state funding as well as no need to sustain this heavy state apparatus, many state organisations (as exemplified by the apex organisations of the three studied cases) undergo changes, i.e. some are dismantled, some run into bankruptcy, some become reorganized, merged and remerged with other institutions. As part of inherited structural path dependency, the apex organisations remain familiar forms of control as during Soviet times, with minor adjustments to the present day agriculture. The apex organisation of the bio-labs, as intermediate state organisations, are not really needed for the present day agriculture, thus they are not supported by the central government. The miserable standing of the Plant Protection makes it find a way for surviving at the expense of its daughter organisations (as it was presented in the cases of non-private bio-labs). The poor status of the Plant Protection organisation, little support from the central government towards Plant Protection organisation and bio-labs signify about little interest of the state to maintain this organisational structure for the present agricultural purposes. The products Plant Protection organisation offers to bio-labs are nothing unique and they can be purchased from the local market or anywhere else. The price of the input is the same as on the local market and the quality is not very high (field notes, interviews with bio-labs, 2010). In addition, as a part of the organisational path dependency, Plant Protection organisation tries to impose its ‚assistance, care and control' over bio-labs as it also used to be part of their work in the past. Therefore, one can say that Plant Protection or the apex organisation of bio-labs imposes its 'cooperation or coordination' towards bio-labs. It is done in order to sustain the organisation during the agrarian change.

The financial status and otioseness of the MTP apex organisation (which is the Regional MTP Union) is very similar to the above described Plant Protection organisation during the agrarian change. Nowadays, even though the MTP Union is the above standing organisation to the MTP, it does not offer any assistance or advantages to the MTP. The apex organisation is left with empty pockets and it is almost dead. Nevertheless the regional MTP Union tries to sustain at any price. The apex organisation of MTP is more a burden than of help to the MTP. The relationships that occur between them are more pushed by the upper organisations, attempting to use their former power, which is, however, not welcomed by the MTP since the apex organisation does not provide any help, but instead is an additional burden for the MTP.

In contrast to those two, the interaction between the F. Co and its apex organisation is very intensive. The relationship is guided by the central state's interests in agriculture and chemical industries, trying to prevent supplying of fertilisers to farmers for free and trying to prevent organisational abuse on the local level. There are also attempts to push the F. Co to work according to the business interests, but the other state interests represented by the state procurement system prevent it.

Thus, AGSOs and other state organisations showed to be a place of persuading personal interests for many actors, especially for the national agricultural policy, organisational members, and state elites and bureaucrats to sustain their living during the transition. AGSOs in Uzbekistan experience a process of differentiation, in an overall context of struggle for institutional survival.

(3) The situation of service providers, similarly to farmers, will not change fundamentally (for the better or worse) as long as the state procurement system is in place.

The reason why many service providers cannot really move or evolve is that they were established to serve collective and state farms during the Soviet times and now continue to serve state agricultural aims (Chapter 3). Therefore AGSOs do not have experience in running the business independently, needless to mention they have never experienced a market environment. Being organized in this particular way the state agricultural system contributes to the wellbeing of the Uzbek centre. In this way it keeps many elements of the Soviet Union system in play, thus the country leadership does not want to change many rules, especially informal ones, which worked during the Soviet times. As a result, the majority of the service providers cannot move ahead, to develop in a market economy.

Agricultural service providers were investigated as a case of interface between the Uzbek central government and the agricultural producers (state---AGSOs---farmers, see section 2.2). In the investigated interface of state---AGSOs---farmers, the state showed to be a very powerful actor influencing the interactions that occur between the service providers (the MTP and the F. Co) and farmers under the procurement system. The methods, mechanisms and rules the central government appeals to are described in the first argument. Service providers and farmers are embedded in the present day agriculture, therefore they have to follow everything dictated by the central government. Also, farmers cannot choose just for growing commercial crops, as it comes just as a compensation for being involved in state procurement system. As we witnessed with some service providers, there are many obstacles put, like limitations in fulfilling extra jobs, by paying debt to the tax office, not giving gas and so on, so they hardly can survive (see Chapter 6 on reinvention by the MTP). Thus, under the current state procurement system farmers are neither motivated in getting higher quality of services nor paying to service providers (in the case of the MTP). Farmers and service providers (the MTP and the F. Co) understand that nothing depends on them, they are just particles in that state regulation and production mechanism: 'Nothing depends on us, we cannot change anything here' (farmers' survey, 2009). Therefore, the majority of AGSOs are not individual decision makers under the present day procurement system in agriculture.

(4) Notwithstanding overwhelming efforts of central state control and regulation, there are dynamics on the ground towards a market economy. During the transition (so far) three types of service providers have emerged.

With the help of the relational typology of agricultural service providers, three types of service providing organisations were identified in present day Uzbekistan. The three case studies provided further details on the distinct characteristics of three types of service providers: state affiliated agricultural service providers, state neglected agricultural service providers and state ignored agricultural service providers (section 8.3).

A first suggested category of agricultural service providers is the state affiliated AGSOs exemplified by the Fertiliser Company (section 3.7.1). This organisation is close to the state as it is an important source of income for the national economy. It is an important input provider for the state procurement system thus the F. Co's functioning is shaped by the state agricultural agenda. There is a frequent interaction between the state administrations with F. Co in regard to the state procurement system. The F. Co allocates a major share of their working time and adjusts daily activities of its organisation to serve state agricultural purposes. The director of the F. Co demonstrated his agency and knowledge to work under the command system, in order to find solutions on how if not prevent, but at least reduce stress related to state procurements system.

Another distinguishing characteristic of the F. Co is that it has an intensive interaction with the state elites and bureaucrats. This interaction involves service providers' abuse by state elites. As I argued, state-affiliated service providers show lower ability to resist state orders and tasks from the bureaucrats, which is strongly embedded into the system of relations from the past. From an inside-organisational point of view, there was more bureaucratic abuse, and overstaffing noticed in the F. Co than in other two service providers.

The interaction between the F. Co and its apex organisation is very intensive. The relationship is guided by the central state's interests in agriculture and chemical industries, trying to prevent supplying of fertilisers to farmers for free. In view of this, the F. Co receives much more financial support from the central government and is strong and developing organisation in contrast to two other studied cases. Uzhimprom (the apex organisation of the F. Co) combines the interests of the state procurement system in agriculture and commercial interests of the factories. One can regard the chemical factories then as a force trying to extricate the F. Co from the web of local games that are skimming of its profits, and from the exigencies of the state procurement system (Van Assche et al., forthcoming). Combining the interests of the factories producing fertilisers and state agricultural purposes stimulate F. Co to prevent giving services for free to the state-ordered farmers.

Reinvention in the F. Co was seen in regard to reduce the shock by the state while mobilising F. Co for state cotton campaign (section 7.2.1). But diversification of organisational activities was driven by the apex organisation. Diversified activities and future plans (starting new activities, undertaking the services provided by weaker organisations like Plant Protection and MTPs, occupying still available niches on the market) by the F. Co prove that it has a solid physical infrastructure, financial basis and state support (section 7.3.1).

The Machine Tractor Park (MTP) is an example of the **state neglected service providers** (section 3.7.2). The MTP represent the organisation type that is moving from the state affiliated to state ignored service providers. The MTP is retained by the state to provide assistance and control in growing cotton and wheat. There is no investment done by the central government in this type of organisation, in contrast to the F. Co. Functioning of MTP is modified according to the state agricultural agenda. Despite strong influence of the state administration upon the MTP functioning, the MTP tires to lobby its interest, employing its informal relations, leadership, diversify its activities, trying to provide services based on money, not employing any kinship or state pressure in order to keep the organisation alive during the transition period. Despite many challenges brought by the transition process to AGSO, the MTP leader has showed the highest level of agency, where the leader puts maximum efforts in order to weaken the link with the state and tries to function as close as possible to the market principles.

During the transition there are not many employment opportunities for the rural population, even being employed in the MTP is already a basis for some people. The MTP is very careful in spending money and thus towards employment in their organisations. In the MTP the director started firing people in order to save the organisation.

The MTP has to provide services to the farmers growing state-ordered- crops. In addition, the MTP is eager to provide services to a variety of clients outside of the state procurement system. In this way, the organisation can secure its survival during the transition period.

The MTP reinvents itself in order to move away from the state influence to survive as an organisation and as part of this process of reinvention develops a substantial degree of localised innovativeness. The MTP managers and primarily director put great efforts to reinvent the organisation through the transition process. In particular the MTP director re-defines the position of the MTP by using his experience gained during the Soviet times to tackle the present problems and to plan the future activities, his connections and kinship relations, his personal qualities rapidly to adjust to the new conditions and to improvise. The MTP has to act or find innovative solutions in the following situations: to prevent bureaucratic abuse of the organisational resource, to find time to do own business apart from assignments in frame of the state procurement system. The director has to cope with a lack of knowledge to work under changing environment; the director has to cope with changes in the organisational structure, where before each manager was responsible for his task and today, the director is responsible for all processes, not much can be done without director's decision; the director has to maintain good relationships with the political power thus with the state administration representatives in the present day procurement system; the director has to use his best personal skills and Soviet knowledge to be able to negotiate with the state administration and clients. So the daily work of the MTP directly depends on the director and his leadership qualities, as well as of the MTP's personnel.

Diversification activities in the MTP include allowing workers to do extra works to earn additional income, designing and producing different caps (nozzles) for the foreign machinery (section 7.3.2). The MTP director appeals to non-financial relationships to settle the financial problems of the MTP. The director with the help of negotiation and bargaining skills manages to reduce the price, while working with banks. The MTP is an example of an organisation where the relationships with the central government are weakening. In order to survive, the MTP remains strategically in cooperation with the state administration. But at the same time, the MTP leader tries to move away from the state influence and tries to develop activities in order to function according to market principles.

Being outside of governmental attention provides a space for manoeuvre for agricultural service providers. **The bio-laboratory (a bio-lab)** served as an example of an organisation that tries to follow market principles while providing services to agricultural producers (section 3.7.3). It is an example of **state ignored service providers**. State ignored service providers showed to be less important to the state agricultural campaign (in line with other services) as well as they are of less financial interest to the state. As a result, they receive less attention from the state, are less bothered by the tasks related to the state agricultural campaign, and hardly receive state support.

Without having state support and financial strength, the organisation is free to decide what and how to act in contrast to the F. Co. Such circumstances of the bio-lab give it space to manoeuvre. It can take individual business-driven decisions; it can develop its business relatively freely from state agrarian politics. Bio-labs are putting all efforts and available materials to develop their business.

The bio-lab reinvents in order to evolve as individual organisation as well as to secure its income (section 7.2.3). The bio-lab's reinvention in the direction of keeping production costs as low as possible under conditions when there is neither state support, nor a solid financial basis. They are usually working for bio-labs instead of hiring employees, and they are usually encouraged to work in return for financial remuneration, like buying some clothes, topping up mobile phone accounts. This is one of the ways by which the bio-lab can reduce some production costs.

The AGSOs in this category are in the position to develop private markets of input supply for fruit and vegetable growers, animal farmers, and peasants who do not produce state ordered crops.

The state ignored service providers are those providers that offer better quality of services and missing ones to all kind of farmers, irrespectively of whether it is for state procurement or commercial farming purposes.

State ignored service providing category encompasses a small share of organisation moving to the market economy. Informal arrangements significantly decline when obtaining a service or a product from the representatives of Western companies as exemplified by Ifoda (section 3.7.4). The market principles are dominant in these organisations. Ifoda accepts only financial relationships regardless of the client. The bio-lab is moving towards this type of relationship with farmers.

8.3 Relational typology of AGSOSs

The role of AGSOs during the ongoing processes of agrarian change was uncovered with the help of the heuristic tool of a **'relational typology of AGSOs'**. The majority of AGSOs are positioned at the interface between state and farmers; they are not individual players but are strongly bound to the state procurement system. The relational typology proved to be a useful device to unravel the characteristics of AGSOs in the period of transition and helped to assess the nature of the established relations as well as dependencies of the three groups of actors (for more details see Chapters 4, 5, and 6). Apart from being adequate as a tool, the typology seems to capture the complexity of the present day conditions in Uzbekistan substantively too. Typology has been enriched through case analysis, which confirmed that the 'hypothesised' types of the heuristic are the 'actual' types in reality.

AGSOs were established during the Soviet times to serve kolkhozes. Nowadays they continue to provide services functional for agriculture, but in particular to the state procurement system. Such circumstances make AGSOs obey the state, without exception. However some differences amongst AGSOs in the ongoing transition period can be drawn. I suggest distinguishing AGSOs into three categories. The relational typology is presented in the following table 3.

Table 3: Relational typology of AGSOs during agrarian change in Uzbekistan

Distinguishing characteristics	**State affiliated service providers**	**State neglected service providers**	**State ignored service providers**
Economic interest of AGSO to the state	Strong economic interest	Low or no economic interest; a heavy burden for the state to maintain this type of organisations	low or no economic interest at all to the state
Importance to the state procurement system	Important input for the state procurement system	Important for the state procurement system; Apart from input, provide knowledge and localized expertise to serve the needs of the current agricultural system	Considered to be less important (in the line with other service organizations) for cotton production

Distinguishing characteristics	State affiliated service providers	State neglected service providers	State ignored service providers
Relation to the state	Unquestionable fulfilment of state's orders required by the present-day procurement system in Uzbekistan	Unquestionable fulfilment of state's orders required by the present-day procurement system in Uzbekistan; however try to avoid this situation	Have nothing or little to do with the state procurement system
	Hardly resistant to the state orders and commands	Try to resist/avoid the state orders and commands	No state orders
Control function in agriculture	Control function over farmers growing state-ordered-crops	Control function over farmers growing state-ordered-crops	Little involved in the state control mechanism of the state-ordered-crops
State support/ indulgences	Receive significant attention and economic indulgences/support from the state	Receive little economic indulgences/support from the state	No economic indulgences/support from the state
Relation to the apex organisation	Stronger link with the apex organisations in contrast to state neglected and state ignored AGSOs	Apex organisations are almost dead and hardly has an impact on present-day functioning	Apex organisations are almost dead and hardly has an impact on present-day functioning
Signs of reinven-	State driven,	Have to please	Innovative, cli-

Distinguishing characteristics	State affiliated service providers	State neglected service providers	State ignored service providers
tion/adaptation to the changes brought by transition	function similarly to the way it was during the Soviet times	the state to survive; in addition, start business activities and provide services to commercial farmers during the night in order to meet the ends	ent driven, custom oriented
Economic situation of AGSOs	Economically strong organisations	former Soviet organisations: economically unsound organizations; reinventing or withering away	Three types: (1) former Soviet organisations: economically unsound organization but with a strong reinvention spirit; (2) established in the recent years private organisations: good economic standing; (3) services supported by the international donor organisations
	Take over the responsibilities of other financially weaker AGSOs	A growing competition from other AGSOs and farmers who launch services previously pro-	Try to provide better quality of services for state-procurement system and

Distinguishing characteristics	State affiliated service providers	State neglected service providers	State ignored service providers
		vided solely by this type of organisation	farmers growing commercial crops, households (where little state attention and thus they offer services in spheres where services are not available)
Inner environment	There is more bureaucratic abuse, overstaffing and less harsh management from state representatives in contrast to state neglected and state ignored AGSOs	Employees are being fired in order to save organisation	'One man theatre'
Level of abuse by local bureaucrats/elites	High level of abuse from local/regional bureaucracy	Abuse from the local level bureaucracy, but AGSOs tries to avoid it	No bureaucratic abuse
Level of freedom from state in decision making	Low	From low to medium	High and develop without state interference
Type of relationships with the state and clients	Same as during the Soviet times	A mix of Soviet and new: Soviet applied to the farmers in state procurement system; new	New type of relationships with state, other AGSOs and clients

Distinguishing characteristics	State affiliated service providers	State neglected service providers	State ignored service providers
		types invented to meet the present day needs of farmers growing commercial crops	

Source: Author's compilation, 2010-2011.

As shown in table 3, the first group is state affiliated service providers (column 2 in table 3). These organisations have strong economic interest for the state and therefore state wants to maintain them under state wing/property. By staying close to the state, these AGSOs continue (like in Soviet times) to unquestionably fulfil state's orders required by the present-day procurement system in Uzbekistan. Moreover, state affiliated service providers get more economic indulgences from the state and are less resistant to the state orders and commands. From an inner-organizational point of view, the closer AGSO to the state, the more bureaucratic abuse, overstaffing and relaxed work attitude were noticed. Thus, this would be the first category of organizations state affiliated agricultural service organisations. Moreover, there is a high level of control over those organizations in contrast to other AGSO. Having a stronger financial basis and governmental support, state affiliated AGSOs take over the responsibilities of other weaker AGSOs, occupy spheres where services are not available, they can render services without prepayment and if they render machinery, it is new (in contrast to outdated machine park of majority of (A)MTPS).

In contrast to state affiliated agricultural service providers, there are AGSOs presenting low or no economic interest at all to the state. These are organsations that fall in the second category of organizations, state ignored agricultural service providers. Likewise, these organisations have a lower direct impact on cotton production (in the line with other service organisations) or have nothing to do with the state procurement system. Thus, the less support they get from the state, the less they are involved in the state control mechanism of the state-ordered crops. A crucial point is that there is low interest of the state in this type of AGSO, as inputs and outputs here are small and not expensive and profits are inconsequential as in state affiliated agricultural service providers. Less attention is allocated by the state to state ignored AGSOs in comparison with state affiliated agricultural service providers. Such status of the state ignored service providers in the agrarian reform offers them a new development path, particularly the opportunity for the development of a private market of services emerges. Non-state producers (commercial farmers, animal husbandry farmers and horticulture farmers), getting less attention from the state, become the primary users of the alternative service providers (markets, shops, small scale import, etc.) mostly due to the fact that the available service infrastructure serves the state-ordered-crop production and thus there is not much foreseen for them by the state. Moreover, some services and inputs they require are missing in the state system. With the development of private markets, new relationships have emerged between service providers and farmers. Besides that, service providers seek new ways of service provision to two types of farmers. State ignored AGSOs offer inputs at the condition of full prepayment, immediate delivery, quality control and full amount. State ignored service providers include organisations that a) were established during the Soviet times, but with the strong desire to develop into private business; b) organisations that already are 100% business driven, working on cash-basis, excluding kinship and provide better quality of service and product; and c) agricultural services supported by international donor organisations.

There is a third type of agricultural service providers that falls between the strongly state affiliated agricultural service providers and state ignored service providers: the state neglected agricultural service providers (column 3 in the table 3). Service providers under this category have the following characteristics: the services provided solely in Soviet times by this type of organisation are available from more sources that have better quality as well as better condition of service provision and assortment nowadays. Thus, a growing competition of similar services from other organisations pushes out the earlier monopolist provider and it is not monopolist any longer. The government does not intend to maintain this organisation as a state entity as the economic situation is poor, most of them run into bankruptcy and require sufficient resources to sustain. Nevertheless, this service remains important to agriculture (in particular for the state procurement system) and thus state control is strongly present in/over these service providers. This kind of organisation is withering away as an organisation, but its service remains vital for state ordered agriculture. Human resources working at such organisations are important for their knowledge of running agriculture under Soviet system and are used as a taskforce to control the present day agriculture. This type of provider is strategically important for state. The main characteristic of these AGSOs are being half-way between becoming private de jure and still strongly subordinated to the state de facto. The state neglected AGSOs are in the process of transition from being state affiliated to state ignored AGSOs; state neglected organisations either wither away with not much business orientation, are unable to cope with the challenges of transition process and the service is carried on by other actors like stronger AGSOs, or they readjust to the present day conditions and sustain.

Apart from being an important device to explore the nature of AGSOs, the relational typology has defined three types of AGSOs that exist in Uzbekistan in the current transition process. According to the characteristics suggested by the typology, we can group the existing AGSOs in Uzbekistan into three categories (see table 4).

In the following table 4, I propose how other service providers not examined in this thesis are categorized according to the relational typology based on their characteristics.

Thus the fertiliser company, fuel supply and banks can be identified as state affiliated agricultural service providers; MTP and AMTP fall in the category of state neglected AGSOs; bio-labs, veterinary services and commodity exchange providers are considered state ignored AGSOs. State neglected AGSOs have an opportunity to provide services and inputs to farmers outside the state procurement system. For example to farmers and rural households dealing with animal production, fruit and vegetable growing, where state service has ceased with the end of the Soviet Union. According to the relational typology of AGSOs, state neglected AGSOs encompass new established private providers like Ifoda, machinery suppliers New Holland and Claas, and supported by the international development organisation (GIZ) bio-labs in Karakalpakstan region of Uzbekistan.

Table 4: Distinguishing existing service providers according to the three categories of the relational typology of AGSOs

Table 4: Distinguishing existing service providers according to the three categories of the relational typology of AGSOs

Service providers	**State affiliated service providers**	**State neglected service providers**	**State ignored service providers**
Fertiliser company	X		
Machine Tractor Parks		X	
Private Bio-lab			X
Semi state/state bio-labs		X	
Alternative Machine Tractor Parks		X	
Commodity produce exchange			X
Veterinary service			X
Fuel supply	X		
Western input suppliers as Ifoda, Claas, New Holland			X
Banks	X		
Cotton gin	X		
Wheat gin	X		
Water Consumer Associations		X	
Farmer's association (information & consulting service)		X	
Missing services for animal husbandries and horticulture			X

Source: Author's compilation, 2012

The findings in this thesis generate development issues to reflect upon that will be of use to other researchers and international development organisations in post-Soviet countries. Based on the author's work and observations as well as communication with researchers from other ex-socialist states, similar types of AGSOs can be found in Ukraine and Bulgaria, equally undergoing change. However, due to little existing research focusing on service providers in agriculture in those countries, it is difficult to expand the application of the developed typology to other countries. Therefore, given the significance of the insight into service provision in rural areas of Uzbekistan after 20 years of agrarian transformation, one possible area for future research would be to look at AGSOs evolvement in other former socialist countries. There is increasing donor interest for instance by the World Bank (Goldberg et al. 2011), Asian Development Bank, UNDP and GIZ in developing extension service systems in these countries, yet there is little knowledge about the current conditions on the ground. Hence, reports such as this are necessary to ensure that donor-developed extension service systems build on the existing but rudimentary forms in an efficient and localised manner, serving the local needs and thus contributing to the wellbeing of rural populations in ex-socialist countries.

8.4 Theoretical preposition

In this section, I summarise my findings in relation to the theoretical prepositions of this thesis. In particular, I will reflect on the role of agricultural service providers in agrarian change in Uzbekistan, Archer's theory of structuration, changing state-producer relationships and different forms of path dependency under the transition period.

Agricultural service providers in this thesis acted as a window into the assessment of everyday forms of agrarian change in Uzbekistan. Studying AGSOs offers a specific perspective on the analysis of agrarian change. AGSOs and their development are a specific aspect of the agricultural change process. The history of the AGSOs as existing in the Soviet context and undergoing change in the last 20 years, is a perfect illustration of the idea of 'structural elaboration' (Archer 1995). Rather than independence being a break or rupture with the past, the process of transformation, since 1990 is characterised as one of slow 'structural elaboration'. There is structural elaboration taking place, showing that things gradually transform and all of the past experience cannot be replayed again, however strongly induced. The present day agriculture of

Uzbekistan mainly rests on the enactment of old rules and roles by the present organisations and leaders involved in agriculture. The structure is **path dependent** that is traced in inherited external (agricultural) structure, organisational structure, and personal or individual structure. By investigating the relationships between AGSOs and state organisations, we could see what has been borrowed from the past experiences in order to fill the gaps of and reinforce the present agricultural system. Due to the cessation of Soviet system supply and support, the central government does not allocate that much support to agricultural producers, AGSOs and state organisations. The structure that is 'elaborated', in episodes of 'reform', trial and error in reorganising agricultural production, and mobilising of local actors, is the Soviet-period structure of state controlled agriculture of which many institutional arrangements remain. The central government employs strict control measures in agriculture, requiring the rural population to grow and collect cotton. The elements of the previous system include the procurement system, a master plan from the government where producers are expected to meet the target production goals, state subsidies, control methods, many meetings, selector, employing state organisation to conduct control in agriculture as well as informal rules of arranging things are also there (Chapter 4). Thus, despite the rigid structure, the system evolves trying to meet the central state's goals under the transition process. The findings show that there are multiple elements of agrarian change enforcing each other and conflicting with each other.

In Soviet times, apex organisations of AGSOs used to be an intermediate level or an interface of the state organisations linking the central government and AGSOs functioning on the local level. The apex organisations used to support functioning and controlled fulfilment of agricultural targets in the case of agriculture. They were managed from the centre in Tashkent where head offices were located and branch offices on the regional and local levels. Since the system was fully state-managed, there was no space for private sector and thus state apparatus was relatively big. But during and due to the transition, apex organisations of many AGSOs are dissolving, as it was exemplified by the cases of MTP Union and Plant Protection organisations. Instead, AGSOs that fell under such apex organisations are now directly manipulated by the state administration (chapters 4, 5, and 6).

Under the present state procurement system in agriculture, **AGSOs are an interface between state and individual farmers**, very strongly influence by the state. The presented data have illustrated that the relationships with the state proved to be very strong and have a strong impact upon the functioning of the AGSOs as well as its impacts upon AGSOs' relationships with farmers and other clients. AGSOs were a part of production process under collectivised agriculture of the Soviet Union. And now, during and due to the transition, the situation is modified in the particular Uzbek environment. To unravel the role of AGSO under the ongoing agrarian change and transition, I looked at relationships established between AGSOs and state organisations (Chapters 4 and 5), AGSOs and farmers (Chapter 6). By looking at the nature of the relationships, we have learned that AGSOs are an integral part of the present day procurement system in agriculture. They depend financially on the state and work in accordance to the Soviet principles, the way they were established in the era of collectivized agriculture. We have learnt that during transition process AGSOs fulfil many other socio-political functions for the central government, local elites and bureaucrats, and for AGSOs' employees. For the central government, AGSOs primarily play a function of control and knowledge carriers for the present day procurement system in agriculture (Chapter 4). For the local elites, bureaucrats and AGSOs' employees, AGSOs provide a space for personal gains and individual survival during the transition (Chapters 5 and 6). Thus, we can regard AGSOs not just as an interface, but as an instrument for the exercising of state control during the ongoing transition period. This position of the agricultural service providers makes them an extreme form of the Technical and Administrative Task Environment concept by Benvenuti (1975) (section 2.2).

AGSOs, being strongly bound to the state procurement system or/and being strongly path dependent, do not have many opportunities to develop. The main reasons for that vary in the three studied cases. The motives for change were presented by the state ignored AGSO, where change is driven by individuals willing to work according to market principles and earn money. They have distanced themselves from the Soviet past and ready to evolve ahead. However, the main obstacle of the state ignored AGSO is its organisational structure, where they used to be fully supported by the Soviet central government. In the present day system, bio-labs are outside of the central state's attention, thus bio-labs hardly get any support from the government in contrast to the state affiliated AGSO. State affiliated AGSOs, who are strongly bound to the state procurement, have less space for manoeuvre, and they are happy under the protection of the state. Changes in the F. Co are driven by the apex organisation of the F. Co. With the support of the Uzhimprom, the F. Co is growing and evolving. There are changes taking place in the state neglected AGSOs, as presented by the case of the MTP. State neglected AGSOs are organisations that are used by the central government for the purposes of the state procurement system, however they are not supported any longer by the central government. MTPs do not get proper support from the central state, the hardware needs to be renewed and they have no finances for that, farmers do not pay for services provided by the MTPs under state procurement. As a result, many MTPs (similarly to state organisation) wither away as they do not manage to sustain themselves through transition or they are not needed by the present day system any longer. Many of such AGSOs are kept to serve state procurement system. The studied MTP is an example of an organisation that wants to move from state affiliated to state ignored AGSOs in order to be more flexible in its activities and decision making, and thus to carry on as an organisation. Based on differences investigated amongst three service providers, with the help of the relational typology, I showed that despite a rigid structure, there are indeed different AGSOs that evolve differently due to the transition process.

By doing research on agricultural service providers, the **nature of an authoritarian state during processes of agrarian change could be studied.** The Uzbek state has different facets and there are different (power) interests in the agrarian change process. There are tension, contradictions and different forces within the Uzbek state. Both can be seen in AGSOs and shape what AGSOs are and how they can develop. It was presented by the example of the apex organisation of the F. Co. On one hand, the Ministry, being part of the state, has to pursue the central government's interest in cotton growing. On the other hand, the Uzhimprom stands for the interests in receiving returns from fertiliser production (Chapters 5 and 6). An empirical instance where the apex organisation assists the regional and local levels of the F. Co to cope with the state abuse of their services is also an example of tensions or contradictions in the Uzbek agrarian transformation process. The tension between prolonging state controlled agriculture and the 'unleashing' of the forces of the market – both of which seem to have support in the state system. Another example of multiple interests of the state administration workers is with local level state representatives. Local elites and bureaucrats have a political and regulating role; they can exert power at their local level beyond their formal administrative mandates. While fulfilling their functional and regulative roles, they at the same time pursue private strategies and interests (Chapters 5 and 6). For example, hokims have acquired more space for manoeuvre in the present system than in Soviet times, but have fewer resources. The implementation of the state order agriculture together with the organisational structure (path dependency) provides a power source for regional governors. They are subordinate to Tashkent's ruling state-ordered agriculture, but this subordination gives them the local power to create their own freedom and extract personal benefits. They need the power, mandate from Tashkent, but they also want to keep Tashkent at a distance for their own purposes - this situation is somewhat different from the Soviet situation. Another example of contradiction in the Uzbek state is state-restricted access to cash in the agricultural economy. By giving little support for the bio-lab's and MTP's development, the central government still expects them to fully support state procurement system. The process of finding one's way through the upheavals of transition was presented in the empirical cases and organisational descriptions (Chapters 3, 4, 5, and 6). The structure of rule in Uzbekistan in my study shows a broad perspective on an authoritarian state and the exercising of its power through the means of agricultural service providers.

Opportunities and constraints of informal relations: While working under the transition environment, informal rules were commonly used in the studied agricultural service organisations. From the case studies, we have learnt that they can be used with different purposes and thus they have shown to have different consequences for the three types of AGSOs. In the Fertiliser Company, informal rules were applied in the cases of employment of the staff members, misuse of organisational resource, fulfilling tasks in regard to the state procurement system, fulfilling tasks of the local elites and bureaucrats. As shown in this thesis, in the F. Co, the informal relations and kin are abusive and applied from the side of the local elites and bureaucrats. However, the Uzhimprom, the apex organisation of the F. Co tries to reduce the use of informal rules, which is abuse in this case. Thus, informal arrangements work until certain extent mostly with the state, semi-state organisation. Thanks to the informal rules that encompass old Soviet rules that are still in play, the Uzbek government mobilises knowledge and resources of the organisations like the F. Co for the purposes of the state procurement system (Chapters 4 and 5).

While running its business, the MTP appealed to informal rules in the cases dealing with the state procurement system in order to make farmers pay or to fasten the transaction with bank for example. As we saw from the example of the MTP, barter and informal arrangements are accepted due to the poor paying capacity of farmers. Farmers can pay in kind (e.g. grains, livestock or externally purchased spare parts for MTP machinery). Also, informal rules were used by the MTP in the cases of working with clients outside the state procurement system, finding innovative solutions for payment with partner organisations and its employees. Similarly to the F. Co, the MTP's resources and staff's knowledge were actively used for the state procurement system despite the contradictions that occurred with the organisation's goals de jure (Chapters 4, 5, 6 and 7). These in-kind payments are then passed on to the staff, but do not contribute to the reduction of the monetary debts of the MTP or show up in the formal accounting system. In situations where there is no cash in the organisation, as presented by the MTP, the director will attempt to instigate informal arrangements or use his negotiation skills or some of his former links in order to find a solution for his organisation and workers. However, the informal ways of receiving services for free from the MTP by the local elites do not bring good consequences to the organisation (Hornidge et al. submitted). Conversely, the informal relations prevent investment in good business opportunities that fit the formal for-profit designation of the MTP. Thus, informal relations compensate institutional shortcomings of the Soviet past as well as the state-ordered system of crop production. Consequently, the spaces for development according to market principles remain limited. Thus in the MTP, informal rules served for the MTP as state pressure reduction and a tool for surviving during the transition process.

In contrast to the F. Co and the MTP, the bio-laboratory appealed to informal rules while trying to find innovative solutions/replacement for the outdated Soviet infrastructure. Informal rules as exemplified by payment in kind with dekhan farms was applied as an alternative payment strategy in the cases of lack of cash during the agrarian change process (Chapter 6).

Since some farmers, due to financial reasons, fail to pay the bio-lab, the director tries to find a solution and to continue cooperation with farmers. Thus, informal payments are accepted from peasants and farmers growing vegetables melons/water-melons, fruits and vegetables. These farmers have poor paying capacity, and also they demand a very small quantity of produce so they are allowed to pay in kind. For instance, peasants can pay by bringing jars, wheat, and sugar. Rural households can bring old books, sugar, and other products, which are needed for the work of the bio-lab. But the director tries to be innovative, and he is looking for ways to move away from using informal rules and to work just according to market principles.

However, there are signs of development according to market principles presented by the examples of the trading of rice and of the trading of spare parts for agricultural machinery, purchasing spare parts in the open-air market and by the case of Ifoda. Private and international organisations operate purely on business relations and do not accept informal ways of payment or doing business.

Members of a large number of state organisations, which apart from their direct responsibilities in the state apparatus, are involved in the state procurement system actively utilise informal rules that encompass Soviet legacies for the purposes of the state procurement system and private goals (Chapters 4 and 5).

Thus, the remaining state apparatus together with informal rules serve as surviving strategies for many state employees under the transition period.

Since there are so many actors from the previous system heavily relying on the rules known from the previous system, it makes it an interesting mix as an input for the present functioning and the future outcomes of agrarian transition in Uzbekistan.

Thus, informal relations not only create opportunities and benefits, but also drawbacks (Wegerich 2004; Oberkircher and Hornidge 2011; Shtaltovna et al. 2011; Van Assche et al. forthcoming; Van Assche et al. forthcoming).

8.5 Policy recommendations for agricultural service providers

AGSOs are an integral part of the present day procurement system in agriculture in Uzbekistan. Yet at the same time their destiny currently seems very unclear. This thesis assessed the specific conditions in Uzbekistan and explored AGSOs with a particular focus on their differences, summarised in the relational typology. Furthermore, I briefly looked at how AGSOs develop in other post-Soviet countries (Chapter 2). Based on the above, two possible future scenarios for the further development of agriculture and agricultural service providers in Uzbekistan seem likely.

1. The state procurement system remains in place. If the state aims at improving the functioning and efficiency of AGSOs, they could return to state control. In this case, state financed investments should increase and communication channels between the state and farmers should be improved, as well as a reform of state institutions should be introduced to boost agricultural production.

2. The state gives up the idea of growing cotton under the state procurement system. In this case, the logic of the majority of present day AGSO's existence will disappear. After that, services required by agricultural producers will be provided by market driven providers, which are likely to evolve out of the previous AGSOs or new service providers (as suggested by typology of AGSOs. See section 8.3). For proper functioning, AGSOs require substantial training of staff on how to run a business under a changing and market-oriented environment, how to deal with financial documentation, do accounting, get a loan from the bank; state incentives (i.e. small subsidies for infrastructure development, decreased taxes); involving donor organisations in strengthening small businesses and private initiatives; establishing bottom-up com-

munication channels and others. The concept of 'integrated service provision' through public-private partnership has shown to be successful elsewhere when of high quality and located close to farmers (Mandel and Humphrey 2002). Linking service provision with extension as it is done in Kyrgystan, Tajikistan and Kazakhstan could be another way to improve agricultural service provision and thus agriculture in Uzbekistan (see section 2.5 for more details).

To function in either scenario, the central state should improve the following:

- The bureaucratic system in Uzbekistan should go through a reorganisation process. Some of the changes needed involve the elimination of bureaucratic abuse of agricultural producers at regional and local levels, professionalization of the bureaucracy, and increasing their payment. Employees of the state should be proud of their work and view their job as a national service. Thus civil service should be rebranded to encourage the best people to join and work for the government.
- Improvement of communication between centre and provinces, that is, the establishment of a proper feedback system between the local level and the centre. The old Soviet methods like selector, conducting numerous checks and meetings for the state procurement system should be improved by reducing unnecessary pressure and increasing efficiency of checks, identifying and solving problems that farmers and AGSOs face while producing for the state procurement system.
- Improvement of old or establishment of new channels of knowledge production and sharing in agriculture. Knowledge is scarce and found in unexpected places, and AGSOs similarly to farmers have to creatively find resources and build connections. They have to invent and adapt to the challenges brought by transition, since no training is provided how to function under a changing environment.
- In terms of organisation restructuring/managerial recommendations: if there is more freedom or desire by the state to improve the situation, there is ample literature and world practice from which to learn this (section 2.4, Mandel and Humphrey 2002). Retraining of present day workers and training of new ones is a must.
- For the improvement of bio-labs: link the innovation brought by development cooperation in Uzbekistan to development of AGSOs. For example, GIZ supports bio-labs in Karakalpakstan in the frame of the GIZ GmbH regional programme "Enhancing Economic and Envi-

ronmental Welfare in the Aral Sea Region (EEWA) in the Karakalpakstan province (Kitching 1998, section 3.6). The suggested innovations do not require special skills, they are simple and non-expensive and can contribute to improvement of the bio-labs where not much state attention is allocated.

REFERENES

Abdullaev, I. and P. P. Mollinga (2010). "Socio-techinical aspects of water management: Emerging Trends at Grass roots level in Uzbekistan." Water (#2): pp. 85-100.

Abdullaev, I., F. Nurmetova, et al. (2008). Socio-Technical Aspects of Water Management in Uzbekistan: Emerging Water Governance Issues at the Grass Root Level. Central Asian Waters: Social, economic, environmental and governance puzzle. M. M. Rahaman and O. Varis. Helsinki, Water & Development Publications - Helsinki University of Technology: 14.

Allina-Pisano, J. (2008). The post-Soviet potemkin village: politics and property rights in the black earth. Cambridge, Cambridge University Press

Archer, M. (1995). Realist Social Theory: the Morphogenetic Approach. Cambridge, Cambridge University Press.

Benvenuti, B. (1975a). "General Systems Theory and Enterpreneurial Autonomy in Farming: Towards a New Feudalism or Towards Democratic Planning?" Sociologia Ruralis Vol. XV(#1/2): pp. 46-46.

Borras, J. (2011). "Agrarian Change and Peasant Studies: Changes, Continuities and Challenges." Journal of Peasant Studies Vol. 36(# 1): pp. 5-31.

Bramel, D. and R. Friend (1981). "Hawthorne, the myth of the docile worker and class bias in psychology " American Psychology Vol. 36(#8): pp. 867-878.

Bremer Cotton Report (2008). www.baumwollboerse.de

Burawoy, M. (1979). "The Anthropology of Industrial Work." Annual Review of Anthropology Vol. 8: pp. 231-266.

Burawoy, M. (2009). The Extended Case Method. Berkeley, University of California Press.

Burawoy, M. and K. Verdery (1999). Uncertain transition: ethnographies of change in the postsocialist world, Rowman & Littlefield Pub Inc.

Chapple, E. D. (1941). "Organization Problems in Industry " Human Organization Vol. 1(#1): pp. 2-9.

Chapple, E. D. (1953). Applied Anthropology in Industry. Anthropology today: an encyclopedic inventory. A. S. Kroeber. Chicago, Univ. of Chicago Press pp. 819-831.

Clapham, C. (1982). Private Patronage and Public Power: Political Clientelism in the Modern State. New York, Palgrave Macmillan

Djanibekov, N., J. P. A. Lamers, et al. (2010). Land consolidation for increasing cotton production in Uzbekistan: also adequate for triggering rural development? Challenges of Education and Innovation for Agricultural Development. K. Labar, M. Petrick and G. Buchenrieder. Halle (Saale), IAMO. Vol. 56: pp. 140-149.

Encyclopedia of Management (2011). http://www.pragmatist.ru/motivaciya-truda/motivaciya-truda-v-sovetskij-period.htmlhttp://www.pragmatist.ru/motivaciya-truda/motivaciya-truda-v-sovetskij-period.html.

Evers, H.-D. and C. Wall (2006). "Knowledge Loss: Managing Local Knowledge in Rural Uzbekistan " ZEF Working Paper Series at Department of Political and Cultural Change Vol. 14: pp. 2-17.

Fox, J. (2007). Accountability Politics: Power and Voice in Rural Mexico Oxford, Oxford University Press.

Giddens, A. (1976). Rules of Sociological Method: a Positive Critique of interpretative Sociologies. London, Hutchinson.

Giddens, A. (1984). The Constitutions of Society: Outline of a Theory of Structuration. Berkeley, University of California Press

GIZ/ZEF (2011). Report on Symposium on Agricultural Service Provision in Newly Independent States of the Former Soviet Union Urgench, GTZ, ZEF/UNESCO, Urgench State University, GFRAS, KRASS: 80.

Goldberg, I., J. G. Goddard, et al. (2011). Igniting Innovations: Rethinking the Role of Government in Emerging Europe and Central Asia. Washington, D. C., World Bank Publications.

Gray, J. (2000). Kazakhstan: a Review of Farm Restructuring. Washington, DC, World Bank Publications.

Hann, C. (2003). The postsocialist agrarian question. Property relations and the rural condition. Münster, LIT Verlag

Hodgson, G. M. (2006). "What are Institutions?" Journal of Economic Issues Vol. 40(# 1): pp. 1-25.

Hofman, I. (2008). The Political economy of rural transformation: Understanding the emergence and differentiation of farm household-enterprises in Khorezm, Uzbekistan,. Wageningen/ Bonn, Wageningen University/ ZEF. Ph.D. Research Proposal: pp. 4-6.

Hornidge, A.-K., L. Oberkircher, et al. (2011). "Reconceptualizing water management in Khorezm, Uzbekistan." Natural Resources Forum Vol. 35(#4): pp. 251-268.

Hornidge, A.-K., M. Ul Hassan, et al. (2011). "Transdisciplinary innovation research in Uzbekistan – one year of 'Follow-the-Innovation'." Development in Practice Vol. 21(#6): pp. 834-847.

Hornidge, A.-K., K. Van Assche, et al. (submitted). Uzbekistan – A Region of World Society: Variants of Differentiation in Agricultural Resources Governance. Soziale Systeme.

Hough, J. and M. Fainsod (1979). How the Soviet Union is governed Cambridge, Harvard University Press.

Humphrey, C. (1998). Marx went away. But Karl stayed behind. Ann Arbor, University of Michigan Press.

Hyden, G., J. Court, et al. (2004). Making Sense of Governance. Empirical Evidence from 16 developing countires Lynne Rienner Publisher, Inc.

ICG (2012). International Crisis Group Tajikistan: Working to Preven Colflict Worldwidc Brussels.

Ilkhamov, A. (2000). "Divided Economy: Kolkhoz System vs Peasant Subsistence Economy in Uzbekistan."

Ilkhamov, A. (2000). "Divided Economy: Kolkhoz System vs. Peasant Subsistence Economy in Uzbekistan." Central Asia Monitor Vol. 4: pp. 5-14.

Ilkhamov, A. (2007). "Neopatrimonialism, interest groups and patronage networks: The impasses of the governance system in Uzbekisan " Central Asian Survey Vol. 26(#1): pp. 65-84.

Information Portal about banks of Uzbekistan (2012). "http://uzbank.net/npacredferm.php."

Ioffe, G., I. Nefedova, J. Zaslavsky (2006). The end of peasantry? The disintegration of rural Russia. Pittsburgh, University of Pittsburgh Press.

Jooshev, P. and N. Mityakova (2008). Extension and Dissemination Strategies in Uzbekistan: An Overview of Issues and Evidences. Bishkek, International Water Management Institute: pp. 1-77.

Jozan, R. (2008). "« État délinquant » ou modèle déviant ? Retour sur le non-respect du traité international de partage de la ressource en eau du Syr Darya." Flux Vol. 2008/1 (# 71): pp. 46-60.

Kandiyoti, D. (2002). How far do analyses of postsocialism travel? The case of Central Asia. Land and Power in Khorezm: Farmers, Communities and the State in Uzbekistan's Decolectivisation Trevisani. Berlin, LIT Verlag: pp. 238–257.

Kandiyoti, D. (2003). "The Cry of Land: Agrarian Reform, Gender and Land Rights in Uzbekisan " Journal of Agrarian Change Vol. 3(Nos. 1 and 2): pp. 225-256.

Kandiyoti, D. (2007). The Cotton Sector in Central Asia: Economic Policy and Development Challenges. The Cotton Sector in Central Asia: Economic Policy and Development Challenges. London, The School of Oriental and African Studies, University of London: pp. 1-238.

Kandiyoti, D. (2007). "Post-Soviet institutional design and the Paradoxes of the 'Uzbek path'." Central Asian Survey Vol. 26(#1): pp. 31-48.

Karimov, I. (1995). Uzbekistan, Along the Road of Deepening Economic Reform. Tashkent University of Michigan.

Kazbekov, J. and A. S. Qureshi (2011). "Agricultural Extension in Central Asia: Existing and Future Needs." IWMI Working Paper Vol. 145: pp. 1-35.

Khan, A. R. (2005). Land system, agriculture and poverty in Uzbekistan. Land, poverty and livelihoods in an era of globalization: perspectives from developing and transition countries. A. H. Akram-Lodhi, Borras, S. M., Kay, C. New York, Routledge. 1: pp. 221-254.

King, A. (2010). "The odd couple: Margaret Archer, Anthony Giddens and British social theory." The British Journal of Sociology Vol. 61(s1): pp. 253-260.

Kitching, G. (1998). "The Revenge of the Peasant? The Collapse of Large Scale Russian Agriculture and the role of the Peasant "Private Plot" in that Collapse." The Journal of Peasant Studies Vol. 26(#1): pp. 43-81.

Kornai, J. (2008). From Socialism to Capitalism: Eight Essays. Budapest, Central European University Press.

Kucherov, S. (1960). "The Future of the Soviet Collective Farm " American Slavic and East European Review Vol. 19(# 2): pp. 180-201.

Lampland, M. (2002). The advantages of being collectivised: cooperative farm managers in the postsocialist economy. Postsocialism. Ideas, ideologies and practices in Eurasia. C. Hann. London, Routledge: pp. 31-56.

LBL, S. C. f. A. E. (2002). The Extension Butterfly. Lindau, Switzerland

Ledeneva, A. V. (2006). How Russia really works: the informal practices that shaped post-Soviet politics and business. Ithaca, Cornell University Press.

Lerman, Z. (2008). Agricultural Development in Uzbekistan: The Effect of Ongoing Reforms. Discussion Papers. H. U. o. Jerusalem. Jerusalem. #7.08: pp. 1-29.

Llewellyn, S. (2007). "Inturducing the Agents..." Organisational Studies (28(02)): pp. 133-153.

Long, N., Ed. (2001). Development Sociology: Actor Perspectives. London, Routledge

Luong, P. J. (2002). Institutional Change and Political Continuity in Post-Soviet Central Asia: Power, Perceptions, and Pacts. Cambridge Cambridge University Press.

Mandel, R. E. and C. Humphrey (2002). Markets and Moralities: ethnographies of postsocialism. London, Berg Publishers.

Markowitz, L. (2008). "Local elites, prokurators and extraction in rural Uzbekistan." Central Asian Survey Vol. 27(# 1): 1-14.

Mendras, M. (2002). The State is Weak, Administrations are Strong. Assessing the Functioning of Bureaucracies in Russia. Honesty and Trust Project workshop Collegium. Budapest: pp. 1-25.

Mielke, K., C. Schetter, et al. (2011). "Dimensions of Social Order: Empirical Fact, Analytical Framework and Boundary Concept." ZEF Working Paper Series(# 78).

Morgounov, A. and L. Zuidema (2001). "The legacy of the Soviet agricultural research system for the republics of Central Asia and the Caucasus." ISNAR report(#22).

Mosiyuk, P., Krysalsky, O., Serdyuk, V, et al. (2001). "Economics and Organisation of Agricultural Service Provision." pp. 1-501.

Niyazmetov, D. (2008). Efficiency of Market Infrastructure Development for Farmers of the Khorezm region. Urgench Urgench State University. Master: pp. 1-125.

Niyazmetov, D., Rudenko, I., Lamers, J. (2012). Mapping and analyzing service provision for supporting agricultural production in Khorezm, Uzbekistan. Cotton, water, salts and soums - economic and ecological restructuring in Khorezm, Uzbekistan. R. I. Martius C, Lamers JPA, Vlek PLG. Dordrecht, Springer Netherlands: pp. 113-126.

North, D. C. (1998). Economic Performance Through Time The New Institutionalism in Sociology. M. C. Brinton and V. Nee. Stanford, Stanford University Press: pp. 247-257.

Oberkircher, L. (2010). Water-Saving in the Landscapes and Lifeworlds of Khorezmian Farmers, Uzbekistan. Institute of Landscape Ecology. Münster, University of Münster. PhD.

Oberkircher, L. (2011). "On Pumps and Paradigms: Water Scarcity and Technology Adoption in Uzbekistan " Society and Natural Resources Vol. 24: pp. 1270-1285.

Oberkircher, L. and A.-K. Hornidge (2011). "“Water Is Life”—Farmer Rationales and Water Saving in Khorezm, Uzbekistan: A Lifeworld Analysis." Rural Sociology Vol. 76(# 3): pp. 394-421.

Oberkircher, L., B. Tischbein, et al. (2010). "Rethinking Water Management in Khorezm, Uzbekistan. Concepts and Recommendations " ZEF Working Paper Series(# 54).
Oi, J. C. (1985). "Communism and Clientelism: Rural Politics in China." World Politics Vol. 37(# 2): pp. 238-266.
Ostrom, E. (1990). Governing the Commons: The Evolutions for Institutions for Collective Actions. Cambridge Cambridge University Press
Paulson, B. (1967). The Role of the Small Intellectual as an Agent of Political Change: Brazil, Italy, and Wisconsin. American Political Science annual meeting. Chicago
Peabody, S., Gurevich L., et al. (2000). Republic of Kazakhstan agricultural postprivatization assistance project: farm privatization and restructuring in Akmola and former Taldy-Korgan oblasts Social assessment and agricultural reform in Central Asia and Turkey. A. Kudat, S. Peabody and C. Keyder. Washington, DC, The World Bank Vol. 461 pp. 173-217.
Pomfret, R. (2000). "Agrarian reform in Uzbekistan: why has the Chinese model failed to deliver?" Economic Development and Cultural Change 48(2): 269-284.
Powell, J. D. (1970). "Peasant Society and Clientelist Politics." The American Political Science Review Vol. 64(# 2): pp. 411-425.
Regional Statistics Office (2010). Information on Machine-Tractor-Parks in the Khorezm region. S. Department. Urgench.
Rumer, B. Z. (1989). Soviet Central Asia: "A tragic experiment". Boston, Unwin Hyman
Sayer, A. (1992). Method in social science: A realist approach. New York, HarperCollins Publishers Ltd.
Sayles, L. (1957). Work Group Behaviour and the Larger Organisation. Research in Industrial Human Relations. C. B. Arensberg, S.; Chalmers, W.E.; Wilensky, H.L.; Worthy, J. C.; Dennis, B.D. New York, Harper & Row: pp. 131-145.
Schwarzman, H. B. (1993). Ethnography in Organizations. New York, Sage Publications.
Sehring, J. (2005). Water User Associations (WUAs) in Kyrgystan: A case study on Institutional Reform in Local Irrigation Management. Zentrum für internationale Entwiklungs-und Umwelt-forschung der Justus-Liebig-Universität Gießen. Gießen, Zentrum für internationale Entwiklungs-und Umwelt-forschung der Justus-Liebig-Universität Gießen.
Serving the People of Central Asia. "www.ec-ifas.org ".

Shreeves, R. (2002). Broadening the Concept of Privatization: Gender and Development in Rural Kazakhstan. Markets and Moralities: Ethnographies of Postsocialism. R. H. Mandel, C. Oxford, Berg pp. 211-235.

Shtaltovna, A. (2007). Role of Governance in the Diversification of Rural Economy: Analysis of Two Ukrainian Cases. Ghent, Ghent University. MSc.

Shtaltovna, A. (2011) Agents of Change: The Role of Agricultural Serivce Organisations in Uzbekistan. ZUR Vol. 24,

Shtaltovna, A., A.-K. Hornidge, et al. (2011). The Re-invention of Agricultural Service Organisations in Uzbekistan – A Machine-Tractor-Park in Khorezm Region. ZEF Working Paper Series. Bonn, Center for Development Research: pp. 1-20.

Spoor, M. (2004). Uzbekistan's Agrarian Transition UNDP.

Spoor, M. (2006). "'Agriculture Reform Policies in Uzbekistan', in: Chandra Suresh Babu and Sandjar Djalalov (Eds.) Policy Reforms and Agriculture Development in Central Asia." Boston: Springer: pp. 181-203.

Spoor, M. (2010). "Agrarian Reform and Transition: What can we learn from 'the East'?".

Spoor, M. (2012). "Agrarian reform and transition: What can we learn from 'the East'?" Journal of Peasant Studies Vol. 39(# 1): pp. 175-194.

State Committee on the Union of Ministers of USSR on the issues of labour and payment (1970). Qualification guidance of employees for office managers and professionals' involved in engineering and economic work in manufacturing plants Moscow.

Strathern, M. (1985). Knowing Power and being Equivocal: Three Melanesian Contexts. Power and Knowledge. Anthropological and Sociological Approaches. R. Fardon. Edinburgh, Scottish Academic Press: pp. 61-86.

Sydow, J., Schreyögg, G, Koch, J. (2005). Paths: Path Dependency and Beyon. 21st EGOS Colloquium. Berlin, Free University of Berlin.

Thelen, K. (1999). "Historical Institutionalism in Comparative Politics." Annual Review of Political Science Vol. 3(# 2): pp. 369-404.

Toleubayev, K. (2009). Plant protection in post-Soviet Kazakhstan: The loss of an ecological perspective. Wageningen, Wageningen University. PhD.

Trevisani, T. (2007). Communities in Transformation: Fermers, Dehqons, and the State in Khorezm. Patterns of Transformation In and Around Uzbekistan. P. T. Sartori, T. . Reggio Emilia, Edizione Diabasis: pp. 185-217.

Trevisani, T. (2008). Land and Power in Khorezm. Farmers, Communities and the State in Uzbekistan's Decollectivization Process. Berlin, Lit- Verlag.

Trevisani, T. (2009). The reshaping of inequality in Uzbekistan: reforms, land and rural incomes The Political Economy of Rural Livelihoods in Transition Economies; Land, Peasants and Rural Poverty in Transition. . M. Spoor. London, Routledge: pp.123-137.

Turaeva-Höhne, R. and A.-K. Hornidge (2012). "From Knowledge Ecology to Innovation Systems: Innovations in the Sphere of Agriculture in Uzbekistan." Innovation: Management, Policy & Practice Vol. 14(# 4): pp. 1819-1843.

University of Texas (2012). "Perry-Castañeda Library Map Collection."

Uzbekistan Agroportal (2010). www.agroportal.uz

Uzbekistan, C. o. M. o. (1997). On Measures of Strengthening the Material and Technical Basis and Increasing Efficiency of Machine-Tractor-Parks. Resolution #152. C. o. M. o. Uzbekistan. Tashkent.

Uzun, V. Y., E.A. Gataulina, V.A. Saraikin, V.F. Bashmachnikov, O.I. Pavlushkina, O.A. Rodionova (2009). "Tendentsii razvitiya i mekhanizmy vzaimodeistviya kruphogo i malogo biznesa v agropromyshlennom komplekse." VIAPI.

Van Assche, K., A. Shtaltovna, et al. (forthcoming). Visible and Invisible Informalities and Institutional Transformation Lessons From Transition Countries: Georgia, Romania, Uzbekistan. Submitted to Interdisciplinary Studies on Central and Eastern Europe.

Van Assche, K., A. Shtaltovna, et al. (forthcoming). "Where did this debt come from? Organizational change, interdependence and role ambiguity in rural Khorezm." Journal of Organisational Studies pp. 1-21.

Veldwisch, G. (2006). "Organisation of Water Distribution." O'zbekiston qikhloq xo'jaligi Vol. 10.

Veldwisch, G. (2007). "Changing patterns of water distribution under the influence of land reforms and simultaneous WUA establishment." Irrigation and Drainage Systems Vol. 21(# 3): pp. 265-276.

Veldwisch, G. (2007). Cotton, Rice, Water: The Transformation of Agrarian Relations, Irrigation Technology and Water Distribution in Khorezm, Uzbekistan. C. f. D. R. Bonn University: 88-108.

Veldwisch, G. (2008). Cotton, Rice, Water: The Transformation of Agrarian Relations, Irrigation Technology and Water Distribution in Khorezm, Uzbekistan. Centre for Development Research. Bonn, Bonn University.

Verdery, K. (2003). The Vanishing Hectare: Property and Value in Postsocialist Transylvania, Cornell University Press.

Verkhovna Rada (2011). "www.rada.gov.ua."

Visser, O. and M. Spoor (2011). "Land grabbing in post-Soviet Eurasia: the world's largest agricultural land reserves at stake." Journal of Peasant Studies: pp. 1-26.

Wall, C. (2006). Knowledge Management in Rural Uzbekistan: Peasant, Project and Post-Socialist perspectives in Khorezm. Center for Development Research. Bonn, University of Bonn. PhD: pp. 1-356.

Wall, C. (2006). Management in Rural Uzbekistan: Peasant, Project and Post-Socialist Perspectives in Khorezm. Center for Development Research. Bonn, University of Bonn PhD.

Wall, C. (2008). Barriers to Technological Change and Agrarian Reform in Khorezm, Uzbekistan. Continuity and Change: Land and Water Use Reforms in Rural Uzbekistan: Socio-Economic and Legal Analysis for the Region Khorezm: pp. 147-163.

Wegerich, K. (2004). "Informal network utilisation and water distribution in two districts in the Khorezm Province, Uzbekistan." Local Environment Vol. 9(# 4): pp. 337-352.

Wegerich, K. (2005). Institutional Change in Water Management at Local and Provincial Level in Uzbekistan. Center for Development Research. Bonn, Bonn University PhD: pp. 1-263.

Wegerich, K. (2006). "'Illicit' water: un-accounted, but paid for. Observations on rent-seeking as causes of drainage floods in the lower Amu Darya Basin." pp. 1-14.

Wegerich, K. (2006). "'A little help from my friend?' Analysis of network links on the meso level in Uzbekistan." Central Asian Survey March–June 2006: pp. 115–128.

Wehrheim, P., Schöller-Schletter, A., Martius, C. (2008). "Continuity and Change: Land and Water Use Reforms in Rural Uzbekistan: Socio-Economic and Legal Analysis for the Region Khorezm. Studies on Agricultural and Food Sector in Central and Eastern Europe." IAMO: pp. 1-203.

World Bank (2011). "World Bank Tajikistan " from: www.worldbank.com/tj.

Yalcin, R. and P. P. Mollinga (2007). "Institutional Transformation in Uzbekistan's Agricultural and Water Resources Administration: The Creation of a New Bureaucracy " ZEF working paper # 22.

Zavgorodnyaya, D. (2006). "Water Users Association in the Republic of Uzbekistan: Theory and Practice." University of Bonn.

Zubets, M. (1996). Agrarian Reform in Ukraine. Kiev.

ANNEX 1: QUESTIONNAIRE FOR THE FARMERS' SURVEY

FARMERS'SURVEY – Questionnaire

Assessment of the farmers' needs regards service provision

(Type of farming)
Name of farm manager: --

Address:
Region: ____
District: ____
Village: ____

Contact info (tel.): --------------------------
Information about your farming practice:

	Parameters	Units	Comments
1	Total area		
	Including irrigated land		
2	Type of crops		
	Cotton		
	Wheat		
	Rice		
3	Livestock		
	Cattle		
	Sheep and goat		
	Chicken		
4	Gardens, Orchards		

1. **Which services do you receive from AGSO?**

Tick 5 the most important agricultural services for your household/business:

	Services	**Which services do you receive?**	**5 most important services**
1.	Water provision (WUAs)		
2.	Machine-Tractor Pools (MTP), Alternative MTP		
3.	Network for the procurement of agricultural production		
4.	Veterinary services		
5.	Insurance services		
6.	Consulting Service		
7.	Outlets for selling oil products (fuel)		
8.	Commercial banks – branches, mini-banks		
9.	Outlets for mineral fertilisers		
10.	Commodity produce exchange		

2. **Are there any services I didn't mention? Yes………..No………..**

If yes, what?

3. **Are there any important services which are not there but you would need them?** Yes………..No………..

If yes, what?

4. **Are there any difficulties you face while working with the service providers?**

Water provision (WUAs)	
Machine-Tractor Pools (MTP), Alternative MTP	
Network for the procurement of agricultural production	
Veterinary services	
Insurance services	
Consulting Service	
Outlets for selling oil products (fuel)	
Commercial banks – branches, mini-banks	
Outlets for mineral fertilisers	
Commodity produce exchange	

5. **How do you manage these difficulties?/ to whom do you appeal to receive help to overcome the problem?**

6. **Are there alternative ways to receive the needed input/services?** Yes………..No………..

If yes, what?

7. **What do you think should be improved in functioning of service providers?**

8. **Have you ever had a chance to receive services from private service providers?** Yes………No…..

If yes,

Do you prefer to work with private and state service providers?

What exactly did you like about it?

9. **Should the services remain state provided or should they be private?**

Water provision (WUAs)	
Machine-Tractor Pools (MTP), Alternative MTP	
Network for the procurement of agricultural production	
Veterinary services	
Insurance services	
Consulting Service	
Outlets for selling oil products (fuel)	
Commercial banks – branches, mini-banks	
Outlets for mineral fertilisers	
Commodity produce exchange	

ANNEX 2: FARMERS RESPONSES IN REGARD TO THE MOST IMPORTANT SERVICES

Responses to the question: please mark 5 the most important services for you

Services	Out of 50 farmers	1st place	2nd place	3rd place	4th place	5th place
1. Water supply (WUA)	30	17	2	4	1	1
2. Machinery (MTPs, AMTPS)	30	8	5	2	3	1
3. Seed supply	22	-	-	2	1	3
4. Veterinary service	30	-	1	-	-	-
5. Insurance	30	-	-	-	-	1
6. Information and consulting	30	1	-	1	-	1
7. Fuel supply	30	1	13	2	3	-
8. Banking services	30	2	-	2	3	1
9. Fertiliser supply (Fertiliser Company)	30	1	8	13	-	1
10. Bio-service (bio laboratory)	22	1	-	-	3	-
11.Commodity produce exchange	30	-	-	1	-	-
12. Sales of agricultural products centres	8	-	-	-	-	-
13.Agricultural produce procurement	8	-	-	-	-	-

Responses to the question: what kind of services do you receive?

Services	How many persons have answered this question?	How many persons have marked this question?
1. Water supply (WUA)	50	45
2. Machinery (MTPs, AMTPS)	50	43
3. Seed supply	42	19
4. Veterinary service	50	36
5. Insurance	50	27
6. Information and consulting	50	28
7. Fuel supply	50	37
8. Banking services	50	26
9. Fertiliser supply (Fertiliser Company)	50	43
10. Bio-service (bio-laboratory)	42	24
11.Commodity produce exchange	50	21
3. Sales of agricultural products centres	8	4
10.Agricultural produce procurement	8	4

ANNEX 3: LIST OF INTERVIEWS CONDUCTED IN UZBEKISTAN AND TAJIKISTAN, 2009-2012

1. Communication with ZEF/UNESCO Urgench employees, 2009-2012;
2. Interviews with the director of the state unitary company MTP 'Agrointechinique' and ZEF/UNESCO project consultant, Urgench, 2009-2010;
3. Interviews with officials at the state committee of de-monopolisation, support of competition and entrepreneurship, Urgench, 2009-2011;
4. Interviews with the head of the AMTP 'Chetkupir' and a farmer at the same time, Urgench, 2009-2010;
5. Interview with the head accountant of the Water Consumer Association in Ashirmat, Khorezm region, June, 2009;
6. Interview with the main specialist at the Department of agriculture at Hokimyat (regional state administration office), Urgench, June, 2009;
7. Three day-visit to the Uzhimprom – Tashkent. Interview with the officials at the department of supply of fertilisers for the state-ordered crop production in Uzbekistan, Tashkent, June, 2010;
8. Interviews with the Chirchiq factory representative to the Khorezm region, Urgench, 2010.
9. Interview with the accountant of the MTP in Shomohulun, Khiva district, Khorezm region, 2009;
10. Interview with the head of Loans' Department at the Agro-bank, Urgench. A number of additional visits to the bank were made while my internship in the Fertiliser Company, 2009-2010;
11. Interview with deputy-director of Urgench Produce commodity exchange, 2009;
12. Interview with the deputy director & a specialist of the Insurance Company Urgench Branch, 2009;

13. Interview with the executive of the Regional Ministry of Agriculture and Water Resources in Urgench, 2009;
14. Interview with the deputy director of the Department of Fuel sales at the Fuel Supply Company, Urgench, 2009;
15. Interview with the Headof the Regional State Veterinary service, Urgench, 2009;
16. Interview with the accountant of the Cotton gin, Shovot district, Khorezm region, 2009;
17. Interview with the main economist of the grain gin 'Shovotdon Mahsulotlari', Khorezm region. 2009.
18. Communication with the director of the NTB centre in Tashkent, Urgench, October, 2010;
19. Communication with the coordinator of Farmers' Association in Tashkent, Urgench, October, 2010;
20. Communication with the senior specialist at the Ministry of Agriculture of Tajikistan, Urgench, October, 2010.
21. Communication with Eshbadalov M. and Jalilov M., AGSOs' experts in Tajikistan, via Internet, October-November, 2011
22. Interview with Muhiddin Nurmatov, USAID extension consultant, Dushanbe, May, 2012;
23. Interview with Mamadshoev A.,an agronomist at Mercy Corps, Shahrituz, 2012;
24. Interivew with the First Deputy Mayor in charge with agriculture, Pengikent, Tajikistan, April 2012;
25. Interview with Nematov I., the director of NGO 'Bakht', Tajikistan, 2012;
26. Interview with Nurmanadov R., the director of the NGO 'Jovid', Dushanbe, Tajikistan, 2012;
27. Interview wtih Ghulomhaydarov S., an agricultural specialist at the NGO 'Shifo', Shahrituz, Tajikistan, 2012.

Interviews and communication with the participants of the GTZ - ZEF/UNESCO Symposium on agricultural service provision in newly independent states of the former Soviet Union, October, 2010, Urgench

1. Interview with Geraedts P., team leader EU SENAS Tajikistan, Cardno UK, Urgench, Ocotober, 2010;
2. Interviews with Isgandarov Y., programme coordinator of Agro Information Centre, Azerbaijan, Urgench, October, 2010;
3. Interview with Nasyrova A., chairperson of management board of Training and Extension System, NGO 'TES centre', Kyrgystan, Urgench, October, 2010;
4. Interview with Yusupbaev J., director of the centre of knowledge dissemination 'Tassay', 2010, Kazakhstan/ The expert at South-Western Research institute of Livestock and Plant Breeding, Kazakhstan, Urgench, October, 2010;
5. Interview with the Suleymanova M., director of Public organisation 'SAS Consulting', Urgench, October, 2010;
6. Interview with Solieva F., Project Manager in Public organisation "SAS Consulting, Urgench, October, 2010;
7. Interview with T. Kalna-Dubinyuk, the Head of Extension Department at the National University of Life and Environmental Sciences of Ukraine, Urgench, October, 2010;
8. Interview with Nesterov I., a product marketing manager, Ukraine and Moldova CNH Services srl, representative office in Kiev, Ukraine. For more information on CNH Services srl see www.cnh.com, online, 2011

ANNEX 4: STRUCTURAL CHANGES IN AGRICULTURAL SERVICE PROVIDERS: FROM SOVIET PERIOD UP TO NOW[44]

Service providers	Collectivised agriculture under Soviet rule (before 1991)	Shirkates period (1997/1998)	"Individualised" farmers and dekhans (2005/2006)	Consolidated agriculture (2008/2009)	Comments
Machinery service	1930s - the MTP was established on the district level and supplied kolkhozes with machinery; 1950s – every kolkhoz had its own machinery park. The MTPs served as repair stations, where machinery from all kolkhozes was brought and repaired. 1960/70s – MTPs were responsible to supply all kolkhozes of the Urgench district with equipment, machinery, spare parts, coal, fertilisers, and seeds. 1980s - the land reclamation department at the MTP was closed down as well as agricultural supply function was moved to other organisations. The MTP continued to deal only with supply of machinery (ploughs, sowing machines, and spare parts) for agricultural purposes.	In 1994 – the MTPs were reorganized into joint stock companies District agricultural machinery service (51% state share); 1997- the MTPs were reorganized into state joint stock companies; 1997 – state subsidized credit for purchasing agricultural machinery (CASE). 1998 – the MTPs were reorganized into state joint stock companies (26% belong to private owners; 25% to the state)	MTP and farmer make a contract. State supportive measures: Between 2007-2010 a decree for the distribution of 25% state shares to other shareholders In this period the demand for machinery from the MTPs declined because machinery is old and worn out. Machinery is offered from other sources like farmers and the fertiliser company	Continuation of the previous period. Private MTPs emerge. Farmers have own machinery; there is a bigger choice of machinery! MTPs offer repair services and render machinery on contract base There is also Uzbekistan agricultural machinery leasing (7-10 years, 7% annuals). 20-30% prepayment (banks can also give credit for this). Rendering services to farmers in frames of the state procurement system: 30-40% prepayment, the rest must be paid within 90 banking days. Farmers rent machinery for their commercial crops at the same conditions as for the state ordered crops Alternative MTPs are neither state nor private organisations at the moment.	Consolidation is not good for MTPs, as farmers become wealthier and they buy machinery and do not appeal to the MTP's services any longer. Machinery exhibitions are taking place.

[44] This table is based on the data collected through the interviews with farmers, agricultural service providers and state administration organizations in the Khorezm province of Uzbekistan (2009-2010)

Service providers	Collectivised agriculture under Soviet rule (before 1991)	Shirkates period (1997/1998)	"Individualised" farmers and dekhans (2005/2006)	Consolidated agriculture (2008/2009)	Comments
Stations for selling livestock	There was 1 centrally provided station per district for selling livestock. There were 12 kolkhozes in the district and 2 of them were involved in animal husbandry in order to supply the rest with live stock			There is one breeding station in the district organised as a joint-stock company.	
Insurance service	There used to be 1 insurance provider in each kolkhoz and the head of kolkhoz was responsible for it.	No changes occurred.	Insurance services began to develop. All types are there: state, private, international	Insurance is not an obligatory service for farmers growing state ordered crops. It is an optional service.	This service is hardly used by farmers.
Info-consulting service or Association of farmers and dekhans	There used to be a trained specialist (agronom) in every kolkhoz.	No changes occurred.	Emergence of farmers' association.	Farmers appeal for information and advice to the heads of the MTPs, former kolkhoz leaders, local representatives of Ministry of Agriculture and Water Resources and to the banks. The Farmers' association seems to exist just on paper; it is located just on the regional level, no local branches exist through the region .	Farmers' association is financed through members' fees; it is supposed to provide consultations on a broad range of issues.

Service providers	Collectivised agriculture under Soviet rule (before 1991)	Shirkates period (1997/1998)	"Individualised" farmers and dekhans (2005/2006)	Consolidated agriculture (2008/2009)	Comments
Fuel supply	The delivery was centralized. There used to be three to four inter-district branches. Kolkhoz received fuel form that inter-district branch.	No significant changes occurred.	Based in shirkats, gas stations were organized. A farmer comes there and makes a contract for delivery of fuel according to the size of farmer's land and technical map. Fuel supply is a state unitary company. It is located on the national, regional, and village levels. In each village there is a branch of the Regional Fuel supplier. In bigger villages there are 2 branches, depending on demand.	No changes occurred.	

Service providers	Collectivised agriculture under Soviet rule (before 1991)	Shirkates period (1997/1998)	"Individualised" farmers and dekhans (2005/2006)	Consolidated agriculture (2008/2009)	Comments
Bank services	There was 1 state bank called – Agrobank. Kolkhoz used to make contracts for delivering fertilisers, machinery, fuel, and the state supplied everything. Kolkhoz did not take part in payment processes. There was only one "account" in kolkhoz and that was it.	No changes Pudrats (groups of dekahns) used to work without accounts, shirkats supplied them with all inputs required for growing cotton. The state supplied with everything as kolkhozes used to do.	Banks are joint-stock companies (with more or less state shares). With the emergence of farmers more banks were established throughout the region. Farmers have their own accounts in the banks and are totally responsible for the financial activities of their businesses. There are options of credits for dekhkans, e.g. for buying livestock and poultry (yearly interest rate- 7%, for 10 years). Overall, the interest rate offered in the banks is 16-18% and it is not often used by farmers because it is too expensive. Additionally, many dekhans are not aware of existing possibilities.	Same as during the previous stage. There are appr. 30 banks in the Khorezm region. With the farmers' consolidation process the local branches of the banks are closing down making it a bit more difficult to access by all rural population. Farmers and AGSOs face a lot of problems while working with banks. However, they learn and improve their work each year. More commercial banks emerging, (Cotton bank, Grain bank and others).	The state's share in banks remains high as long as the 3% subsidized credit for agriculture is there. 'Banks are trouble-makers, they are not used to offer services yet. The reform in there is needed. Banks also need to make profit. But the public doesn't use them often yet and banks can't really make profit. It will come with time' (interview with the former state official, May, 2010)

Service providers	Collectivised agriculture under Soviet rule (before 1991)	Shirkates period (1997/1998)	"Individualised" farmers and dekhans (2005/2006)	Consolidated agriculture (2008/2009)	Comments
Fertilisers (F. Co)	The F. Co used to deliver fertilisers to kolkhozes according to the size of the land.	No changes	The F. Co makes a contract with farmers for delivering fertilisers, indicating the schedule of delivery and the amount. 2001- The F. Co was united with the Plant protection; 2005 - Fertiliser's company and Plant protection were put apart. Fertiliser's company was transferred under the supervision of State Joint Stock Company "Uzkomyosanoat'. Plant protection - to the Ministry of Water and Agriculture.	51% state shares in the F. Co Evolvement of payment by farmers to the Fertiliser's company under the state procurement system: 2005 - 35%, within the next 60 days the remaining money to be paid. 2006 - 50% 2007 - 75% 2008-2010 - 100% The private market of services is emerging. Within the past 5 years, the number of private shops has increased from 5 to 70 in Khorezm. The main users of these shops are farmers with commercial purposes.	What will happen to the F. Co? - it will be there. It will focus on agro-chemical services - application of suspension, carriage of faeces. More perspectives for the future - broaden the spectrum of fertilisers for fodder crops, for animal husbandry, horticulture. We plan to have a hang glider in each district for fertilisers' application from the air; opening shops with different necessary things for the subsidiary plots and for farmers (interview with the official of the F. Co, 2010)

Service providers	Collectivised agriculture under Soviet rule (before 1991)	Shirkates period (1997/1998)	"Individualised" farmers and dekhans (2005/2006)	Consolidated agriculture (2008/2009)	Comments
Plant protection/ Bio-laboratory	Regional and district plant protection branches provided services to the kolhozs	No changes 2001- The F. Co was united with the Plant protection. Before 2004 plant protection was supplied with honey, grains, sugar for production of the insects.	2005 – The F. Co and the Plant protection were put apart. Fertiliser's company was transferred to the coordination of Chemical Production. Plant protection – under the Ministry of Water and Agriculture. After reorganisation, it became a small organisation and its financial status is very poor.	Same. Bio-service, bio-labs are available in all districts. Chemical methods are less in use nowadays. Bio-labs are NGOs, private entities, join-stock companies with either 35 or 51% state shares, belonging to the plant protection branches and to the Ministry of water and agriculture. There are 93 bio-labs with state shares and 16 private bio-labs in the Khorezm region. The functions of the plant protection branches were partially transferred to the Fertiliser's company (defoliation) due to its poor performance since 2008. Pesticides, herbicides are available from private small firms and on the market	It is not in a good state at the moment (regional/district plant protection organisations are bankrupt). Plant protection can be re-united again with the Fertiliser company, as the latter ones partially perform their functions and work financially sustainable) and are under the supervision of Ministry of Chemical production or turned into private, joint-stock companies (as bio-labs evolve). In any case, plant protection should be improved as it is an important organisation to agriculture.

Service providers	Collectivised agriculture under Soviet rule (before 1991)	Shirkates period (1997/1998)	"Individualised" farmers and dekhans (2005/2006)	Consolidated agriculture (2008/2009)	Comments
Fruit and vegetable processing, Storage facilities	The fruit and vegetable companies used to set up contracts for fruits and vegetables with the kolkhozes.		Agro firms (self-financing) and Green World (joint venture, soon – self-financing) – collect fruits and vegetables from farmers and dekhans. Dekhkans usually sell fruits/vegetables on the market. Some bigger companies interested in processing of fruits or vegetables contact the regional administration. The former ones link with the company with available and interested fruit/vegetable farmers.	Same as during the previous period.	
Commodity produce exchange	No commodity exchanges	No commodity exchanges	State+ self-financed Commodity exchanges. It appeared in 2004-2005, along with the idea to provide all necessary inputs for commercial farming for farmers and dekkhans. Almost everything can be purchased here at the condition of full prepayment	The same. Commodity exchange also provides the missing inputs for the farmers growing state ordered-crops. There are 2 places to play in Khorezm – Urgench and Honki.	

Service providers	Collectivised agriculture under Soviet rule (before 1991)	Shirkates period (1997/1998)	"Individualised" farmers and dekhans (2005/2006)	Consolidated agriculture (2008/2009)	Comments
Veterinary	In kolkhoz times there used to be 1 veterinarian and vet stations in each kolkhoz	No changes	State veterinary service is combining self-financing activities of the local vet stations. State providers injections for all animals are for free in case an epidemic emerges. It also checks the quality of livestock on the markets. State covers their salaries and public utilities of the district/local branches, other costs are covered by the entrepreneurial activities.	There are state and private vet services in each village (depending on demand).	
Source: Author's compilation (2011)					

ANNEX 5: AGRICULTURAL PRODUCTION IN KHOREZM PROVINCE, 2008-2009

	All producers			Agricultural enterprises			Farmers			Dekhan farms		
	2009	2008	+, -	2009	2008	+, -	2009	2008	+, -	2009	2008	+, -
Total arable land	220616	210715	9901	2172	3251	-1079	179730	168890	10840	38714	38574	140
Wheat	52062	48769	3293	357	301	56	35529	31627	3902	16176	16841	-665
Corn	1650	1789	-139	49	67	-18	1123	1249	-126	478	473	5
Rice	14375	8848	5527	327	136	191	12214	7228	4986	1834	1484	350
Other cereals	1199	1316	-117	239	287	-48	600	461	139	360	568	-208
Total cereals	69286	60722	8564	972	791	181	49466	40565	8901	18848	19366	-518
Cotton	95185	100725	-5540	597	1705	-1108	94588	99020	-4432			
Industrial crops	2084	1652	432	47	151	-104	1977	1458	519	60	43	17
Potato	4197	4147	50		2	-2	690	773	-83	3507	3372	135
Carrot	12647	11801	846	19	27	-8	4340	4262	78	8288	7512	776
Melons and gourds	5011	4505	506	67	29	38	2960	2499	461	1984	1977	7
Seeds for melons and gourds	12	8	4	12	8	4						
Fodder, total	32194	27155	5039	458	538	-80	25709	20313	5396	6027	6304	-277

Source: Regional Statistic bureau, Khorezm province, Uzbekistan (2009)

ANNEX 6: RECOMMENDATION FOR USE OF ALLOCATED SUBSIDISED LOAN TO FARMERS FOR COVERING EXPANSES FOR GROWING COTTON, 2008-2009

№	Types of expenses	Expenses of growing 1 ton of raw cotton		The sum of subsidised loan	January	February	March	April	May	June	July	August
		sum	%									
1	Salary	38241			3337	5550	5942	5500	5564	5628	3340	3340
2	Fertilisers	85938			189	8641	23011	12350	16347	9676	7862	7862
	including:											
	Nitrogen	61973			189	1561	6126	12350	16347	9676	7862	7862
	with phosphor	21182			0	5400	15782	0	0	0	0	0
	with potassium	2783			0	1680	1103	0	0	0	0	0
3	Plant Protection	7075			600	982	1366	982	982	984	589	590
	a) Chemical method	3538			300	491	683	491	491	492	295	295
	b) Biological method	3537			300	491	683	491	491	492	294	295
4	Seeds	5555			800	1400	1700	1655	0	0	0	0
5	Machinery service (MTP, AMTP and organisation that have machinery) and leasing expenses	22450			1055	2800	5499	3218	3760	2376	1871	1871
6	WUA services	2942			180	605	672	611	286	267	160	161
7	Fuel supply	36222			1100	3800	10402	5200	5366	4132	3111	3111
8	Expenses for polyethylene film	338			0	70	140	128	0	0	0	0
9	Expanses for electricity	3591			0	648	1050	669	383	383	229	229
10	Single tax for land	3780			0	787	787	500	550	525	316	315

11	Other expenses (expenses in the process of production of cotton and allocations for storage)	5068			1034	1040	1033	1311	203	204	122	121
	TOTAL:	211200	100%		8335	26323	51602	32124	33441	24175	17600	17600

The first deputy of the Minister of Agriculture and Water Resources

U. Barnoev

The chairman of the Fund on the agricultural goods purchased for the state needs:

Y. Khidirov

Fund for agricultural accounts, purchasing for the state needs under the Ministry of Finance of the Republic of Uzbekistan

700078, Tashkent, Mustakillik square, 5 tel: (371) 139 10 62, 139 48 52

January 28, 2008 No.10/1-31

To the Commercial banks (according to the list), to the main Agencies of the Central bank, to the regional inspectors of the Fund,

according to the point 6 of the thesis registered in the state list under No.1675 by the Justice on April 14, 2007,

by the Ministry of Finance with the collaboration of Ministry of Agriculture and Water Resources

"The recommendations on the ultimate amount of credits for financing of each type of 2008 cotton growing expenses"

were prepared and directing for using loans in agricultural enterprises.

We deliver these recommendations to the necessary regional branches and filiations of the bank in the short period of time.

The director of Fund: Y. Khidirov

RECOMMENDATION FOR USE OF ALLOCATED SUBSIDISED LOAN TO FARMERS FOR COVERING EXPANSES FOR GROWING COTTON, 2009

№	Types of expenses	Expenses of growing 1 ton of raw cotton		The sum of subsidised loan	January	February	March	April	May	June	July	August
		sum	%									
1	Salary	81700	19	49020	4085	3472	12868	6537	6944	6944	4517	3653
2	Fertilisers	172000	40	103200	8050	7025	25875	12790	15442	15442	7356	10720
	including:											
	Nitrogen	120400	70	72240	0	0	9990	12790	15442	15442	7356	10720
	with phosphor	44720	26	26832	5431	5723	15678	0	0	0	0	0
	with potassium	6880	4	4128	2619	1302	207	0	0	0	0	0
3	Plant Protection	17200	4	10320	860	725	2715	1376	1462	1462	588	1132
	a) Chemical method	8600	50	5160	430	363	1357	688	731	731	294	566
	b) Biological method	8600	50	5160	430	362	1358	688	731	731	294	566
4	Seeds	11180	2,6	6708	1069	1219	1742	2678	0	2009	0	0
5	Machinery service (MTP, AMTP and organisation that have machinery).	43000	10	25800	2150	1882	6718	3440	3655	3655	1520	2780
6	WUA services	7740	1,8	4644	387	338	1210	620	657	658	228	546
7	Fuel supply	74820	17,4	44892	3644	3030	12545	5238	6563	6562	1920	5390

8	Expenses for polyethylene film	860	0,2	516	180	180	156					
9	Expenses for electricity	9030	2,1	5418	452	395	1411	722	767	768	261	642
10	Single tax for land	7310	1,7	4386	365	320	1142	585	622	621	203	528
11	Other expenses (expanses in the process of production of cotton and allocations for storage)	5160	1,2	3096	258	226	806	414	438	438	107	409
	TOTAL:	430000	100	258000	21500	18812	67188	34400	36550	36550	17200	25800

The chairman assistant of department on economical reforms of the Ministry of agriculture and water resources: B. Nurmikhammedov.

The chairman of the Fund on the agricultural goods purchased for the state needs: Y. Khidirov.

THE NORM OF MONTHLY EXPENSES FOR PRODUCING RAW COTTON, 2010

№	Types of expenses	Expenses of growing 1 ton of raw cotton		The sum of subsidised loan	January	February	March	April	May	June	July	August
		sum	%									
1	Salary	85434	17	51260	4272	3738	13348	6078	4557	10369	4856	4032
2	Fertilisers	185945	37	111567	5244	7057	22647	18836	27941	8765	9459	11618
3	Plant Protection	19096	3,8	11458	955	835	2984	1270	1048	2456	1008	902
	a) Chemical method	9548	50	5729	478	417	1492	635	524	1228	504	451
	b) Biological method	9548	50	5729	477	418	1492	635	524	1228	504	451
4	Seeds	23620	4,7	14172	2581	2112	9479	0	0	0	0	0
5	Machinery services	61814	12,3	35897	2548	2548	9813	4011	2982	8108	3917	1970
	MTF and AMTP services	37088	60	22253	2244	1529	5888	2407	1789	4864	2350	1182
	Leasing fees	24725	40	14836	1496	1019	3925	1604	1193	3244	1567	788
6	WUA services	6030	1,2	3618	301	264	942	463	521	524	318	285
7	Fuel supply	95485	19	57292	6401	3970	14840	6734	5297	10431	5785	3834
8	Expenses for polyethylene film	2512	0,5	1507	502	502	503	0	0	0	0	0
9	Expenses for electricity	6533	1,3	3919	327	286	1224	303	363	763	345	308
10	Single tax for land	7036	1,4	4221	352	307	1303	342	407	807	371	332
11	AWA fees	5025	1	3015	251	319	786	387	233	636	265	238

12	Other expanses	4020	0,8	2412	201	176	628	309	250	446	212	190
	TOTAL:	502549	100	301530	25127	22014	78497	38733	43599	43305	26546	23709
	Cumulative				25127	47141	125638	16437 1	20797 0	251275	277821	301530

The chairman assistant of department on economical reforms of the Ministry of agriculture and water resources: B. Nurmikhammedov.

The chairman of the Fund on the agricultural goods purchased for the state needs: Khidirov.